ECOLOGY OF EVERYDAY LIFE

ECOLOGY OF EVERYDAY LIFE

Rethinking the Desire for Nature

Chaia Heller

Montréal/New York
London

Although changed significantly, a version of Chapter One appeared in *Ecofemnism:
Women, Animals, Nature*, Greta Gaard (Temple University Press: 1993)

Black Rose Books No. CC275
Hardcover ISBN: 1-55164-133-X (bound)
Paperback ISBN: 1-55164-132-1 (pbk.)

Canadian Cataloguing in Publication Data

Heller, Chaia
Ecology of everyday life : rethinking the desire for nature

Includes bibliographical references and index.
Hardcover ISBN: 1-55164-133-X(bound)
Paperback ISBN: 1-55164-132-1 (pbk.)

1. Human ecology. 2. Environmental degradation—Social aspects.
3. Nature—Social aspects. I. Title.

GF21.H44 1998   304.2'8   C98-901072-4

Cover Design by Associés libres, Montréal

BLACK
ROSE
BOOKS

C.P. 1258	2250 Military Road	99 Wallis Road
Succ. Place du Parc	Tonawanda, New York	London, E9 5LN
Montréal, H2W 2R3	14150	England
Canada	USA	UK

To order books in North America: (phone) 1-800-565-9523 (fax) 1-800-221-9985
In Europe: (phone) 0181-986-4854 (fax) 0181-533-5821

Our Web Site address: http://www.web.net/blackrosebooks

A publication of the Institute of Policy Alternatives of Montréal (IPAM)
Printed in Canada

CONTENTS

For my parents, Audrey and Bill
whose encouragement, generosity, and love of life
are a gift, an inspiration.

ACKNOWLEDGEMENTS

If it takes a village to raise a child, it takes several communities to write a book. Writing is indeed, a social process and this book would never have been written without the support and insight of a wide number of wonderful people.

I give great thanks to the Institute for Social Ecology for providing a forum in which to develop these ideas since I first arrived there fifteen years ago. To Dan Chodorkoff who welcomed me into the ISE at such a young age, allowing me to develop my abilities as a teacher, and I thank Peggy Luhrs and Ynestra King for leading me to into the world of feminist theory and practice. I am grateful to Paula Emery, for her wit, wisdom, and relentless sense of style, and to Claudia Bagiackas, Betsy Chodorkoff, Cathy Chodorkoff, and Michael Mazzenga, whose everyday labors of intelligence and love make the Institute for Social Ecology a much needed reality. I am grateful to the entire faculty at the ISE who are working to keep revolutionary ideas alive, and I thank Cindy Milstein and Janet Biehl for their enduring integrity and commitment to keeping the struggle honest and visionary. I owe many thanks as well to Zoë Erwin who has always made me feel that my ideas were worth getting out into the world. The students in my ecofeminism classes over the years have been central to writing this book. I am deeply appreciative of all of their questions and criticism that have continually challenged me to push my ideas forward. I owe much gratitude to Brian Tokar who has been a tremendous source of support and inspiration both as a fellow thinker and as a friend, helping me through this process at difficult times, and reminding me of the reasons to keep going.

I also thank the wonderful people at the University of Massachusetts Department of Anthropology who have inspired me to venture into unfamiliar and exciting new theoretical waters. Jackie Urla, Brooke Thomas, Ralph Faulkingham, Art Keene, and Rick Fantasia are just some of the people who have been an invaluable source of support and guidance. I thank Arturo Escobar for his generous support for my work and for making my return to school nothing short of a joy.

For reading and making valuable comments on parts of this book, I thank the many women of Northampton who gave their time and attention to this manuscript;

special thanks to Janet Aalfs, Sally Bellerose, Susan Stinson, Allison Smith, Susan Edelstien, Elena Deutch, and Hillary Sloin. To the folks in the living-room study group of '96, I give great thanks to you all: Zoë Erwin, Eric Toensmeir, Tania Tolchin, Rebecca De Witt, Jonathan Stevens, and Nancy Lustgarten, for receiving my ideas with intelligent criticism and generous enthusiasm—and a special thanks to Bob Spivey for endless moral support, encouragement, and willingness to laugh and listen. Big thanks too, to James Creedon and Morgan Kennedy who speed-read their way through the book in (one of) its last incarnations, giving important insights and suggestions.

Thanks also to the friends who helped me through the last pangs of producing this manuscript. Thanks to Brian Tokar, Cindy Milstein, Peter Staudenmaier, and in particular, Greta Gaard who painstakingly took care to make sure the book entered the world in good form. I thank Greta for the years of encouragement and wisdom which have motivated and inspired me. Thanks also to Carolyn Merchant for the time and generosity it takes to support the arrival of a new book.

Friends and family have given me strength that has made this book possible. I am indebted to Ilana Boss-Markowitz, Allison De Fren, and Nancy Bael for decades (!) of love, humor, and endurance, for being crucial touch stones that make life navigable. I thank Alison Prine for reminding me always of courage and imagination, and of course, poetry. Crow and Clove have provided me with years of patience, advice, and wisdom. Hillary Mullins, I thank for being my first feminist buddy and Jaimé Morton, I thank for music and a solid first try. To my newer friends in the Valley, Beverly, Bob and Sam Naidus-Spivoy, Nancy Lustgarten, Valija Ivalds, Lisa Beskin and Robin, Allison Smith, Sally and Cindy, Susan Stinson, and Elena Deutch, I thank you all for providing me with sustenance and phone-friendship during periods when I could not leave my computer. And to my incredible family, thanks to my parents who have patiently supported every page of this journey, my sisters Laura, Carol, and my brother-in-law Jorge, who keep me anchored in love and humor from year to year. Thanks to Allen, Judy, Jane, and Paul Kronick whom continue to provide generosity and warmth. And I wish to express deepest appreciation to Sandra and Dick Smith for providing endless encouragement, insight, and support to pursue the work and life I love.

Thanks to Steve Chase who first put the bug in my ear, Pavlos Stavropoulos and Riki Mathews at Aigis who got the ball rolling, and I owe many, many thanks to Dimitrios Roussopoulos and Linda Barton of Black Rose Books who were able to get the book out into the world.

I am most grateful to Murray Bookchin whose relentless vision, bravery, and brilliance has blazed a trail upon which I have traveled, and hope to continue to travel, for many years to come. I am reminded daily of my good fortune to have found in Bookchin a mentor, generous and encouraging, who has been willing and able to gently prod me to recognize my own potential to think, write, and speak about revolutionary ideas. For this, and for so much more, I will always be indebted.

This book would not have been written without Lizzie Donahue, leprechaun extraordinaire, whose love, encouragement, humor, and flair for the absurd, gave me the peace of mind to finally sit down and write.

INTRODUCTION

ECOLOGY AND DESIRE

Ecology is as much about desire as it is about need. While the ecology movement of the sixties addressed the need for clean air and water for survival, it also expressed a popular desire for an improved quality of life. People took to the streets in the seventies to fight nuclear power; but many also took to the land to build ecological communities hoping to enrich their social relationship as well as their ties to the natural world. Ecology addresses two demands, then—one quantitative, the other qualitative. Born out of the call for enough clean water, air, and land to survive, ecology is also the demand for a particular quality of life worth living.

Ecology And The Dialectic Of Need And Desire

As political protest to ecological degradation began to wane in the mid-eighties, an emphasis on quality of life issues held steady. Enthusiasm for nature-based spirituality, as well as for natural foods and medicine, reflected a continuing popular desire for health and meaning associated with ecology. However, this emphasis on quality of life has taken on an individualistic tone often expressed through personal changes in life-style and consumption habits. If middle-class North Americans feel socially disempowered to ensure the planet's survival, they can at least command the buying power to ensure that their individual lives will be ecologically pleasurable in the short term.

In turn, ecology has taken on a romantic dimension. For privileged peoples within industrialized capitalist contexts, there is a tendency to desire a 'pure' or 'innocent' nature that is prior to or outside of society. Such ecological discussion can range from a longing to protect an ideal 'mother nature', to a yearning to return to a golden age that may have never existed. The growing popularity of wilderness exploration trips on the one hand reflects a genuine wish for a meaningful connection with the rest of nature. But on the other hand, such ventures echo the myth of the romantic hero strutting off into the 'wilds of nature', turning away from the society he has left behind.

More and more, questions of desire upstage questions of need within ecological discussion. Insulated from (and often desensitized to) the immediate effects of ecological breakdown, people of privilege still have sufficient natural resources to survive. However, not everyone is protected from immediate ecological crises. Due to the effects of capitalism, racism, sexism, and state power, most people on the planet are obliged to design a very different ecological agenda. While also sharing the desire for quality of life, most of the world's people are increasingly under pressure to emphasize questions of need and survival in their work for ecological justice.

There exists a global 'division of ecological labor' in which, while the poor in the Southern hemisphere are forced to work to sustain the *viability* of life, addressing questions of access to food, water, and land, many in the North are able to work to establish a *quality* of life, considering what *kind* of food to eat, what quality of water to drink, as well as what kind of spiritual or cultural sensibility to embrace. Again, while *all* people desire a better quality of life, the question of who has the freedom to fulfill these desires is largely informed by global questions of power and privilege.

And yet, this division of privilege cannot be reduced to geography. Due to the global nature of advanced capitalism, there is a bit of the North in the South and a bit of the South in the North. Indeed, as the under-class swells within the U.S. and Europe, a privileged élite continues to grow within the Southern continents as well. Still, despite these complexities, it makes sense to point to this global division: it allows us to acknowledge conditions of inequality under global capitalism that are generally manifested on opposite sides of the equator.

In response to this global division of ecological labor, many well-meaning activists suggest that we should eliminate 'superfluous' qualitative questions to focus on issues of survival alone. Concerned with the ecological 'bottom line', they reduce ecology to quantitative issues of demographics and population, calculating the number of people that may survive in ecosystems without exceeding a 'carrying capacity'. Or, romanticizing the predicaments of indigenous peoples, activists of privilege often reduce these struggles to questions of need and subsistence, perpetuating the myth of the 'needy primitive' who depends on the benevolent assistance of white men.

When activists focus solely on questions of ecological need and survival, they fail to recognize the qualitative concerns of poor peoples who also share desires for a meaningful and pleasurable quality of life. In this way, they ignore the fact that most poor people cannot access the things they may desire. A vast number of people in the U.S. cannot afford quality organic produce enjoyed by middle and upper-class peoples, nor can they afford the time, cost, equipment, or transportation to take pleasure in the vistas of 'nature' by vacationing in national parks—no matter how much they might like to.

Each community, rich or poor, has its own struggle for quality of life. Activists in Harlem fight for a clean and beautiful neighborhood park for their children to enjoy, while also organizing campaigns for clean air. In turn, intrinsic to indigenous struggles for ecological sustainability are attempts to protect meaningful cultural practices that are also threatened by capital-driven poverty and ecological devastation.

By reducing the ecological agenda of others to issues of need, ecological activists miss the opportunity to redirect their own desire for an ecological quality of life in a more radical direction. In fact, the desire for an ecological way of life among both poor and privileged peoples carries within it the nascent demand for an ecological society, a demand that has potentially revolutionary implications. For, once we collectively translate this desire into political terms, we are able to challenge a global system that immiserates most of the world's inhabitants, forcing them to forgo their desires, lowering their ecological expectations to the level of mere survival. Keeping a desire-focus within the ecology movement keeps our demand for satisfaction, vitality, and meaning alive, invigorating our ability to envision a socially and ecologically desirable society.

What is more, a needs-focused agenda directs our attention away from the qualitative dimensions of everyday life that are so crucial to ecology. Ecological activists need not repeat the same errors committed by the old left which emphasized issues of quantitative need over matters of qualitative desire. Marx believed that a universal condition of material need caused all social strife and injustice. Accordingly, Marx asserted that after material inequity was abolished through the revolutionary process, social relations would be automatically improved, restoring quality of life to realms outside of labor as well. Marx could not have anticipated the degree to which capitalism would invade and erode the realm of home and the everyday in the post-war era. Again, for Marx, it was primarily the sphere of work that was poisoned with alienation, and it was there that he placed the locus of his theory.

The sixties brought a needed challenge to Marxist theory. Groups such as the Situationists in France, as well as sectors of the American New Left expanded their focus to address the encroachment of capitalism into everyday life. The New Left's emphasis on such qualitative domains as sensuality, art, and nature stood as a response to Emma Goldman's apocryphal warning to Marxists decades before: "If I can't dance, in your revolution, I'm not coming." As these movements illustrated, a focus on desire keeps our eyes on the qualitative dimension of life. It allows us to attend to the ways in which the process of commodification extends into our relationships with each other and with the natural world, reducing parents to 'child-care providers', the sick to 'consumers of health-care', and nature to patentable 'genetic material'. A focus on desire offers us a way to counter this emptiness with a desire for a qualitatively new world of our own making.

Finally, focusing solely on need and survival naturalizes conditions of ecological scarcity and destruction. When we lose sight of the qualitative dimensions of life, we lose the ability to contrast the world that *is* to the world that *ought* to be. We lose the ability to see and name the very institutions that prevent society from becoming the desirable creation that it ought to become. Paradoxically, focusing on desire allows us to expose the social mechanisms that produce conditions of scarcity. Such a focus reveals the true solution to the ecological division of labor: to challenge the political and economic institutions that force the world's majority to struggle to satisfy basic ecological and social needs. Clearly, this challenge would entail a politicization of an ecology predicated on a redefinition of need and desire as well as a transformation of economic and political power. Not only would we have to rethink the quality of our needs and desires, but we would have to explore new ways to meet them within new social and political institutions.

Nature And Desire: Toward A New Understanding

As the contemporary ecology movement approaches the end of its third decade, the ecological division of labor remains intact. What impedes ecology from fulfilling its potential to transform institutions that fabricate social and ecological need in the first place? Certainly, a primary cause of the ecological division of labor is a global hierarchical system of political and economic power which benefits the privileged who, in turn, keep the system in place. Yet, in addition to this problem of social hierarchy, there is also a crucial issue regarding how privileged peoples within advanced capitalist society frame concepts of nature and desire.

Ideas about nature and desire stem from centuries of ideology that support existing political and economic structures in the West. To a large extent, we inherit our romantic ideas regarding nature from thinkers of the colonial era. By the eighteenth century, Rousseau became the first in the West to position the category of nature in explicit moral opposition to society, describing nature as an exotic, eden-like state of innocence to which 'man' must emulate. Indeed, the nature we know and love in the West is largely born out of the colonial imagination. It is Diderot's Tahiti where the colonizer fixed his gaze upon an exotic other dwelling in an objectified realm of purity.

We have also inherited a Germanic understanding of nature formalized during the nineteenth century by thinkers such as Ernst Haeckel. For Haeckel, who coined the term 'ecology' in 1867, nature represented a pristine and mystical realm bound to the people of the German nation, a wholesome haven which must be protected from exogenous elements. We in the West are the inheritors of such understandings. Our notions of nature are often abstract and romantic, proscribing idealized places and times to protect or return to, rather than proposing radical social change that could provide the basis for a free and ecological society.

Our ideas regarding desire are also highly problematic. As citizens of a liberal capitalist society, our desires constitute an amalgam of individualistic, competitive, and acquisitive yearnings. Consequently, we tend to see ourselves as individuals destined to compete for scarce resources, striving to fulfill a range of personal desires for sex, wealth, status, or security. Desire is largely viewed as a matter of self-interest expressed within the realms of work, politics, and even love. Informed by a capitalist sensibility, desire is often reduced to yearnings for an accumulation of private property, both material and symbolic. Even matters of spirituality, meaning, and aesthetics tend to be translated into quests to 'acquire' personal truth and beauty. Rarely do we view desire as a yearning to enhance a social whole greater than our selves, a desire to enrich the larger community.

When such approaches to nature and desire meet, they give rise to an unfortunate approach to ecology. Combining an individualized and capitalistic notion of desire with an abstract and romanticized understanding of nature, we engender a movement of people who long to return to a more pristine quality of life by consuming artifacts and experiences that they deem 'natural'. Ecology becomes a movement of people who see themselves as individuals and consumers yearning for ecological asylum rather than as part of a social whole that strives to radically transform systems of power.

Thus, our ideas of nature and desire direct ecological criticism away from social change and toward the protection of a 'nature' to be enjoyed by privileged peoples. This tendency has dismayed social change activists who regard middle-class desires for wilderness preservation and personal life-style as being insensitive to the needs and desires of poor people.

Yet as we have seen, the question is not whether to focus on ecologically-related need *or* desire; clearly, we must address both. The question is what *kind* of desire will inform the movement and what *kind* of 'nature' will be the subject of that desire within ecological discussions? Will it be an individualistic desire for a nature that is understood to be outside of society? Or will it be a *social desire*, a yearning to be part of a greater collectivity that will challenge the structure of society to create a cooperative and ecological world?

I believe social ecology, feminism, and social anarchism can help illuminate a definition of desire that is profoundly social, rather than purely romantic or individualistic. This is crucial because, while our society offers us a variety of ways to describe the many dimensions of individualistic desire, we are offered a paltry vocabulary with which to describe a social understanding of desire. We are saturated by consumerist rhetoric of 'personal satisfaction', yet rarely do we hear eloquent discussion regarding the cooperative impulse, or regarding the craving for a free and non-hierarchical society. Instead, our society worships at the fountain of capitalism whose insatiable waters of material greed and sexual domination crowd out the opportunity to cultivate a desire to regenerate rather than deplete cooperative social and ecological relationships.

Yet while there is little talk of social desire within the domain of liberal capitalism, it continues to speak its own name within many social movements. Within social anarchist movements of the Old Left and the more recent movements of the New Left, there exists an implicit understanding of both the complex needs and desires which people bring to the revolutionary project. Activists in the civil rights, women's liberation, gay and lesbian liberation, ecology, and anti-war movements fight to recreate social life from a qualitative perspective in addition to opposing material inequality in society.

Indeed, the feminist and ecological movements are compelling illustrations of 'desirous movements.' Radical feminists of the sixties and seventies demanded more than to merely survive male violence and sexual inequality: they also addressed a wide spectrum of aesthetic, sexual, and relational concerns. Similarly, the ecology movement of the seventies and early eighties wanted more than to stem ecological destruction. The back-to-the-land movement crystallized a desire for a more healthful and sensual expression of everyday life.

In turn, the civil rights movement embodied a sensual impulse in its plea for 'brotherhood' between the races expressed in Martin Luther King's speech, "I Have a Dream". King's speech represents one of the most passionate and poetic in history, giving voice to the collective desire of the African American community not just for political and economic equality, but for a particular quality of life infused with dignity, beauty, and cultural integrity. Civil rights activists sought to awaken a sensibility based on mutual respect and a reclamation of collective cultural self-love.

Even within movements driven primarily by material scarcity, a dimension of desire plays a vital role. Among the anarchists in the Spanish Civil War were peasants who fought not merely for an allotment of bread, but for a spectrum of social and moral freedoms as well. What made their struggle different from communist sectors within the Old Left was their demand for beauty, pleasure, and collectivity as well as access to food, land, and control of the means of production. Film footage of this revolution reveals the dual nature of the struggle: while revolutionaries risked their lives in combat, they also, in the process, converted luxury hotels previously owned by the rich into halls in which everyone could eat, drink, dance—and enjoy, if for only a moment, the quality of life for which they were willing to die.

This book represents an attempt to begin to rethink our notions of desire in the hope of radicalizing our approach to ecological questions. It emerges out of the belief that ecology should not be reduced solely to issues of physical need and survival, but should also embrace the desire for an improved quality of everyday life that can only be achieved through a profound transformation of social, economic, and political institutions. It also represents an attempt to reconsider our understandings of nature by challenging romantic and dualistic assumptions that underlie notions of what constitutes ecological change.

The Ecology of Everyday Life brings together some of the ideas I have grappled with during the years 1984 to 1998. These chapters were written from within the movements in which I traveled as an activist and a teacher; movements ranging from the greens and ecofeminist movements to the anarchist movements that have re-emerged in recent years. The ideas presented here were developed during a time in which activists in these movements were rethinking such basic categories as nature, desire, identity, and politics, reaching for more nuanced and complex understandings of questions of power related to social and ecological questions.

These ideas also emerged from my work as a psychotherapist and social worker. For over a decade, I worked with a range of people—poor and privileged—developing an appreciation for the everyday struggles that people endure as they search for meaning, community, and pleasure in a world that is often alienating and disempowering. Through this work, I began to understand the enormous burdens and joys that people bring to ecology; I began to appreciate both the personal and political sources of their hopes and dreams for a better world.

Coming of age in a greater-New York suburb in the seventies, and raised in a conservative middle-class Jewish family, my own voyage to feminism, social ecology, and social anarchism has been complicated indeed. The 'nature' I knew was an acre of woods behind my elementary school, 'politics' was Richard Nixon and the cold war, and 'feminism' was the white business-woman standing proud with her briefcase on the cover of *Ms.* magazine.

This book reflects my attempt to understand the origins of my own dreams and assumptions about society and nature, as well as my ongoing struggle to articulate new ways of thinking about social and ecological change.

The 'radical ecologists' I address and critique in these chapters are my friends, fellow activists, students—and myself, as I, too, continue to work to transcend the epistemological and institutional constraints this society imposes upon a world we are all trying so desperately to transform.

Throughout the eighties and nineties, I recognized a need for privileged people active within such movements to be more critical about the way they approach ecological issues. Focusing on the trials and tribulations within the radical ecology movement, the chapters in part one were written in an attempt to encourage others in the movement to consider the historical and political forces that lead their ecological activism in a romantic or individualistic direction. These chapters treat ecology as a discussion that is constrained by systems of racism, capitalism, sexism, and state power; a discussion in which activists must locate themselves in reference to questions of social privilege and power.

I wrote the middle set of chapters in an effort to expand our current vocabulary for discussing desire within progressive movements. Dismayed by what I saw as a reduction of desire to romantic and individualistic terms, I decided to explore the cooperative impulse within social anarchism, feminism, and social ecology to uncover a more 'social' expression of desire that I believe draws out a cooperative sensibility within ecological discussion. The second chapter in the section is an exercise in thinking through what it means to be sensual, creative, and dynamic, appealing to the metaphor of the 'erotic' to point to different facets of social desire. I wrote this chapter in response to a tendency among radical ecologists to counterpose questions of intuition and reason or spirituality and rationality. I wanted to explore the possibility of transcending this dualism by using a different metaphor for conveying deeply meaningful social and ecological experiences that are marked by both emotion and rationality.

Finally, the last section brings together the idea of social desire with a new understanding of nature drawn from social ecology. Positing desire as social, and nature as 'natural evolution', I explore a 'social desire for nature': a desire to create cooperative social and political structures to establish a society that allows people to participate constructively in natural evolution. To ground an ethics for a 'social desire for nature', I look to Bookchin's natural philosophy, concluding that a rational desire for nature entails the decision to create an ecological society based on direct democracy. Finally, I explore a framework for thinking through how to enact such a social desire for nature, illustrating a way to reflect a broad political and revolutionary vision within particular ecological and social struggles.

My purpose is to be both critical and reconstructive, illustrating limitations in our ecological thinking while offering insight into how to transcend those constraints by creating a more radical understanding of both nature and desire. I have come to believe that it is crucial for society to become aware of the ways in which ecological ideas are informed by qualitative questions of desire and longing, a desire that must be approached in a social rather than individualistic direction if true political transformation is to occur.

To challenge previous ecological thinking is not merely a matter of arguing that the approaches taken by radical ecologists have been politically biased or socially constructed. What is necessary is not to criticize previous thinking for being a product of history, but to understand the historical processes which have produced such thinking in order to create new ways of conceptualizing ecological change. A critical discussion of 'ecological thinking' is particularly crucial today because, as I have just mentioned, a major tendency in the U.S. ecology movement has been to polarize questions of reason and emotion so that ecological yearning for such ideas as 'wilderness,' 'community', or animal liberation are often understood as lying outside the domain of rational reflection and discourse. Too often, ecology has become a thing to 'feel' rather than a thing to 'think' as well.

In this book, I have tried to transcend this binary between thinking and feeling to create an understanding of 'informed desire'. I believe that we do not degrade the integrity of our desires, be they spiritual or aesthetic, by understanding their origins and implications. I also believe that our thinking is of little value if our thoughts do not move us to take compassionate and political action to improve the lives of other people and of the planet. Ultimately, I believe that a desire informed by an appreciation of history, politics, and ethics can help us to look critically and passionately at how to solve the social and ecological problems that we face today.

Of the many thinkers I have read, there are four who, for me, most exemplify the ability to synthesize reason and passion. For each of these thinkers, there is one work that inspired me to write this book: first, *Post-Scarcity Anarchism* by Murray Bookchin; second, an essay written by Audre Lorde called "The Uses of the Erotic: The Erotic as Power"; third, the chapter "The First Bond" in Jessica Benjamin's book *The Bonds of Love;* and fourth, a short poetic essay by James Baldwin entitled "The Creative Process." [1]

I point to these pieces as a way to illustrate the sources of a few of the many threads I have knitted together in an attempt to develop a new understanding of the 'desire for nature'. I am teetering on the shoulders of these great thinkers—one a natural and political philosopher, one a feminist poet and theorist, another a feminist psychoanalytic theorist, and yet another, a novelist and essayist—trying to perhaps bring together pieces of myself that I can in turn, integrate toward a new understanding of the questions I pose in this text. As a poet, psychologist, social ecologist, and feminist, I have tried over the years to consider the social and political conditions that are necessary to allow *all* people to express their desire or creativity in ways that will make the world a more interesting, ethical, and pleasurable place.

I offer this book as a reflection on how to draw from a variety of sources, both reasoned and impassioned, to think about how to create a more desirable and ecological world. It is my belief that desire fleshes out the revolutionary project, inciting us to expect more than that which we need, enlivening us to demand the fullness of social and ecological life, in all of its passionate complexity.

Notes

1. I am indebted to these writers for inspiration and direction. While I have drawn inspiration from many of these writers' works, the pieces mentioned here represent for me paticularly important sources for new ways of thinking about desire. See Murray Bookchin, *Post-Scarcity Anarchism* (1969; reprinted by Montreal: Black Rose Books, 1986); Audre Lorde, "Uses of the Erotic: The Erotic as Power," in *Sister Outsider* (New York: The Crossing Press, 1984); Jessica Benjamin, "The First Bond," in *The Bonds of Love: Psychoanalysis, Feminsm, and the Problem of Domination* (New York: Pantheon Books, 1988); James Baldwin, "The Creative Process," in *The Price of the Ticket: Collected Nonfiction 1948-1985* (New York: St. Martins, 1985). I cannot help but include a quote from this last essay of Baldwin, who I hope, would forgive me for modifying the pronouns:

> Societies never know it, but the war of an artist with society is a lover's war, and the artist does, at best, what lover's do, which is to reveal the beloved to themself and, with that revelation, to make freedom real.

PART I

THE DESIRE FOR NATURE

CHAPTER ONE

RESCUING LADY NATURE: ECOLOGY AND THE CULT OF THE ROMANTIC

Ecological awareness of the planet peaked in 1972 when astronauts first photographed the planet, revealing thick furrows of smog encasing a blue and green ball. 'The world is dying', became the common cry as the planet, personified as 'Mother Earth', captured national, sentimental attention. Nature became rendered as a victimized woman, a Madonna-like angel to be idealized, protected, and 'saved' from society's inability to restrain itself. Decades later, we still witness popular expressions of the desire to protect 'nature'. As we observe each April on Earth Day, politicians, corporate agents, and environmentalists take their annual leap into the romantic, ecological drama, becoming 'eco-knights' ready to save helpless 'lady nature' from the dragon of human irresponsibility.

The cult of romantic love, which emerged first in the twelfth century poetry of the French troubadours of Longuedoc, still provides a cauldron of images and metaphors for today's depictions of nature.[1] Contemporary Western representations of 'mother nature' emerged out of this "cult of the romantic" tradition based on a dialectic between an heroic savior and an ideal lover. Indeed, the metaphors and myths used to discuss ecological problems often find their origins within romantic literature. Yet despite its association with love, romanticism often shows its cool side when it surfaces within ecological discourse. While often expressing a desire to protect 'mother nature', it may ignore the social and political struggles of marginalized peoples. In particular, romantic ecology fails to challenge the ideologies and institutions of social domination that legitimize social injustice. Instead of challenging institutions and ideologies of domination within society in general, romantic ecology too often points its sword toward abstract dragons such as 'human nature', 'technology', or 'western civilization', all of which are held responsible for slaying "Lady Nature." In turn, romantic ecology often veils a theme of animosity toward marginalized groups under a silk cloak of idealism, protection, and a promise of self-constraint. It not only refuses to make social liberation a priority, but in some cases, actually holds the oppressed responsible for the destruction of the natural world.

Before exploring the romanticization of nature, we might look briefly at the romanticization of women in the middle-ages as depicted in romantic love poetry. Unlike 'modern romance' which consists of moon-lit dinners, crimson sunsets, and sexual contact, medieval romanticism represents an unconsummated love. As in the story of Tristan and Iseult, an Arthurian romance in which two ill-fated young lovers spend their short lives in pursuit of an unconsummated, yet passionate love, lovers rarely express their desire for each other physically. [2] Instead, classical romance emphasizes the act of passionate longing, an intensity of feeling that is heightened by deprivation. Knightly and courtly romance is a love from afar, expressing its desire in the form of passionate love poetry.

The origins of romantic love may be traced to Plato's concept of desire.[3] Platonic love emerges out of metaphysical dualism which divides the world into two discrete material and spiritual domains. The realm of spirit, or 'idea', is regarded as superior to the transient and perishable realm of the body, or matter. According to Plato, intellectual and sexual knowledge is most valuable when gleaned independent of physical experience for ideal love represents a disembodied yearning that remains 'unpolluted' by physical contact. For Plato, the highest form of love is the intellectual 'fondling' of eternal, rational ideas found in geometry, philosophy, and logic. For the romantic, ideal love is the exercise of sexual restraint and an intellectual expression of passion through love poetry.

IdEAlizATiON, PROTECTiON, ANd CONSTRAiNT

Romantic poetry often consists of the wistful desire of a man for an idealized woman to whom he rarely gains sexual access. This 'noblest desire' thrives in a realm of purity, in contrast to marriage, which is seen as merely reproductive. Courtly romance consists of elaborate rituals of devotion in which the lover promises to protect the beloved from human and mythical villains, while also promising to restrain his sexual desire for the beloved lady.

However, the lover's inauthentic idealization of his beloved is reflected in the incongruity between the celebratory spirit of the poetry and the actual social context in which it was written. Certainly, the idealized, pedestaled position of the women in the poetry does not reflect the actual status of the majority of women in feudal society. The theme of romantic protection represents a fantastical projection by the male romantic. Even when the lady's lack of social power seeps through into the fabric of the poetry, her powerlessness is framed as a need for knightly protection. The romantic fantasizes that the woman needs knightly protection from predators instead of recognizing her desire for social potency. The simultaneous act of elevating and protecting the idealized woman in romanticism allows the hero to sustain the fantasy of the woman-on-pedestal while indirectly acknowledging her very real low social status. In this way, the romantic becomes the protector of the pedestaled woman, creating a subtle amalgamation of male fantasy and social reality.

The fantasy of romantic protection is predicated on the lover's promise of sexual self-constraint toward his lady. However, romanticism never questions the social conditions which make such constraint necessary. A romantic story would lose its charm if the knight were to challenge the social or political institutions which render the Lady powerless in the first place. Romanticism patently accepts that men inherently desire to plunder women, while regarding promises of male self-control as heroic acts of self-mastery.

At this juncture, we might ask why the romantic fails to critique the social conditions which regard idealization, protection, and male self-constraint as a necessary good? Surely, the lover wishes his beloved to be truly free. Perhaps the function of romantic love is to camouflage the lover's complicity in perpetuating the domination of the beloved. Perhaps idealizing, protecting, and promising to constrain the desire to 'defile' the beloved emerges out of a power structure from which the lover knowingly or unknowingly benefits and thus wishes to maintain. In the name of protecting the beloved from the dragon that threatens to slay her, then, the knight actually slays his beloved himself: He slays 'His Lady's' self-determination and agency in the world. In this way, the knight is really the dragon in drag.

Romance, Hierarchy And Alienated Desire

In addition to prescribing idealization, protection, and self-constraint, romanticism also prescribes an alienated form of desire and knowledge. Romantic love is based on the lover's desires, rather than on an authentic knowledge of the beloved. The romantic's love depends on his fantasy of his beloved as inherently powerless and good according to his definition. He views his beloved through a narrow lens, focusing only on a minute, vulnerable section of her full identity; meanwhile, the rest of her body becomes a screen for the projection of his fantasy of the ideal woman. The romantic glosses over information about his beloved which contradicts his personal yearnings. In this way, romantic love is a form of reductionism, reducing the idea of 'woman' from a full range of human potential to a tiny list of male desires.

Romanticism is a way of knowing which is wedded to ignorance. The romantic clearly does not know his lady to be a woman capable of self-determination and resistance. He does not recognize her ability to express what is most human, including her capacity for rationality and critical self-consciousness. Most significantly, the romantic is unaware of women's capacity for self-assertion through sabotage and resistance.[4] The subject of romantic poetry rarely includes stories of 'good' women poisoning their romantic lover's food, or stories of admirable women being emotionally unavailable to their lovers. Few are the poems or stories which tell of strong, lovable women resisting compulsory motherhood, marriage, and yes, even heterosexual romance. The cult of the romantic erases the idea that woman can be a wrench in the machine of male domination.

Romantic love represents an attempt to love and know another from behind a wall of domination. Indeed, true love and understanding can only occur when both subjects are free to express their own desires. The knight can only love the lady if he is willing to relinquish his power over her, supporting her struggle if and when she requests it; then and only then, can they begin to talk about love.

Romantic desire is predicated on a hierarchical separation between the lover and the beloved, separations that are, in turn, predicated on hierarchies based on such factors as sex, age, race, and class. Traditionally, just as the master may romanticize the slave, men may romanticize women, adults may romanticize children, and the rich may romanticize the poor. These separations are reinforced by institutions and ideologies that exaggerate differences between identity groups within social hierarchies. In turn, while the idea of gender is polarized and performed through rigid gender roles and children are segregated in school-ghettos, adults are ghettoized in work places often segregated by race, class, and sex. These structural barriers facilitate the condition of social alienation based on ignorance. Romantic desire flourishes between the walls of social hierarchy as the privileged paint their own romantic fantasies of the lives and condition of the oppressed. When all is said and done, the privileged know very little about the history and lives of those upon whose backs their privilege weighs.

Contemporary Ecology And The Romantic Protection Of Nature

Today, society's increasingly alienated understanding of 'nature' opens the way for romantic discussions of ecology. More and more, the 'nature' we know is a romantic presentation of an exaggerated 'hypernature' marketing researchers believe we would be likely to buy. The less we know about rural life, for instance, the more we desire it. Ideas of 'nature', a blend of notions of exotic 'wilderness' and 'country living', form a repository for dreams of a desirable quality of life. So many of us long wistfully for a life we have never lived but hope to find someday on vacation at a Disneyfied 'jungle safari' or glittering sweetly inside a bottle of Vermont Made maple syrup.

Murray Bookchin, creator of the theory of social ecology, said years ago that the more the rural dissolves into poverty, development, and agribusiness, the more we would see romantic images of the rural in the media.[5] Sure enough, in the 1990s, just as the family farm crisis peaked, commercials and magazine ads were suddenly riddled with rural images: Grandfathers were everywhere, rocking on rustic porches, uttering wise platitudes regarding the goodness of oat-bran. Red-cheeked kids began running down dirt roads after a day of hard wholesome play in the country, ready for Stove-Top Stuffing. And just as the Vermont family dairy farm began to vanish in the early eighties, "Ben and Jerry" bought the rights to the Woody Jackson cow graphic, transforming the Holstein cow logo into the sacred calf of Vermont.

The tendency to idealize nature is often accompanied by the desire to protect a 'nature' that is portrayed as weak and vulnerable. Each year on Earth Day, an epidemic of tee-shirts hits the stores depicting sentimental images of 'nature'. One shirt in particular presents an image of a white man's hands cradling a soft bluish ball of earth. Huddled around the protective hands, stands a lovable crowd of characteristically wide-eyed, long-lashed, feminine looking deer, seals, and birds. Under the picture, written in a child-like scrawl, reads the caption, "Love Your Mother." The message is clear: nature is ideal, chaste, and helpless as a baby girl. We must save 'her' from the dragon of 'every man'.

Ironically, this romantic posture toward nature often promotes an uncompassionate portrayal of the causes of 'nature's woes'. The desire to protect nature often conceals the underlying desire to control and denigrate marginalized peoples. For example, during the late 1980s, members of several radical ecology groups were called to task for attributing environmental problems to over-population and immigration. The *Earth First!* journal has consistently over the years advertised a sticker that reads "Love Your Mother, Don't Become One." Paradoxically, the same radical ecologists who express a romantic desire for 'Mother Earth', also suggest that mothers themselves are to blame for the denigration of nature. In the name of 'protecting mother earth', Third World women are reduced to masses of faceless bodies devouring the scarce resources of the world. Meanwhile Gaia, the idealized mother herself, sits elevated on her galactic pedestal awaiting knightly protection from women's insatiable wombs.

The fantasy of romantic protection blends perceptions of social reality with desire and fantasy. The romantic can remain disdainful and ignorant of systems of social oppression while pursuing the desire to protect 'Mother nature'. However, removing the veil of romantic protection from population debates reveals population imbalances to be the result of a continuing legacy of patriarchy, colonialism, racism, and capitalism. For centuries, while suppressing indigenous cultural practices that regulate fertility, social and political forces have created economic and cultural demands for increased fertility. Throughout history, small scale cultures have been able to control population through a range of medicinal, technical, and sexual practices ranging from post-natal sexual taboos to herbal abortificants.[6] However, as capitalist wage economies emerged throughout Europe and the now Third World, factors of poverty, high infant mortality, and religious reproductive control unsettled cultural practices that balance reproduction. Indeed, factors including lack of reproductive health care, colonially induced religious taboos against contraception, high infant mortality, poverty, and families, needs for child labor within cash economies create a context in which women bear more children than they historically would have otherwise.

Moreover, population fetishists rarely highlight the fact that 'overpopulation' in the Third World contributes little to the overall depletion of the earth's resources. While one middle-class person in the U.S. consumes three-hundred times the food and energy mass of one Third World person, First World corporations and the U.S. military are the biggest resource consumers and polluters. In 1992, with less than 5 percent of the world's population, the U.S. consumes 25 percent of the world's commercial energy.[7] As Bookchin stated as early as 1969, there is something disturbing about the fact that population growth is given the primacy in the ecological crisis by a nation which has a fraction of the world's population and wastefully destroys more than fifty percent of the world's resources.[8] Consistently, those who consume the most are held the least accountable while the poorest are blamed for the world's problems. Meanwhile the real corporate and state perpetrators of ecocide remain hidden under a shroud of innocence. Statistical numbers games that calculate national resource consumption to include a woman on welfare as well as that of General Motors, or people of color as well as whites, create an illusion of a generically 'human' consumer. Such games serve to focus on numbers and demographics rather than social relationships and institutions such as capitalism.

Deep ecologists such as Bill Devall and George Sessions have also often failed to address the social conditions of poor women. While their writings express a desire to protect 'nature', their romantic approach to ecological problems often entails a less than compassionate analysis of the origins of and solutions to the denigration of nature:

> Humans are valued more highly individually and collectively than is the endangered species. Excessive human intervention in natural process has led other species to near-extinction. For deep ecologists, the balance has long been tipped in favor of humans. Now we must shift the balance back to protect the habitat of other species...Protection of wilderness is imperative.[9]

A careful analysis of this quote reveals the sexism and racism which often underlies a desire to protect 'nature'. Constructing an unmediated category of 'humanity', these writers hold an abstract 'human' responsible for the destruction of nature. However, it is unclear just whom is subsumed under this category of 'human'. Do the authors refer to disenfranchised peoples who, rather than participating intentionally and profitably in "human intervention" over nature, are degraded along with natural processes themselves?

Blaming 'humanity' for nature's woes blames the human victims as well as perpetrators of the ecological crisis. Certainly, those most victimized by capitalist processes are not to blame for ecological destruction. For example, due to structural adjustment programs, laborers in so called Third World countries are coerced by multi-national conglomerates and international development agencies to become instruments of ecological destruction.

In the attempt to repay debt to the World Bank, local communities throughout the Third World are forced to convert land areas to cash-cropping sites, destroying ecosystems that have sustained them for centuries.[10] Poor workers in both the First and Third Worlds fight daily to survive the low-pay slavery which subjects them to toxic and deadening working conditions—yet they too, are subsumed under the general category of the accountable 'human'. Failing to expose the social hierarchies within the category of 'human' erases the dignity and struggle of those who are reduced to and degraded along with 'nature'. But again, the liberation struggles of marginalized peoples are never quite so romantic as the plight of the ecological activist struggling to protect 'nature'.

Ecology And The Desire For Purity

Romantic ecology is often predicated on the desire for purity. This desire carries within it a yearning to destroy all that is corrupt within society, as well as that which threatens the integrity of 'nature'. Choosing their own dragon of choice to bear the blame for ecological corruption, each yearns for a romanticized time, place, and people of the past whom they deem as having been idyllic. For some, it is 'the foreigner' who destroys the integrity of a race, morality, or culture that the romantic craves so bitterly. For others the dragon is identified as 'modernity' whose technologies, cities, and 'progressive' ideas degrade a past social order that is romanticized as having been morally and ecologically superior. What purists share in common though, is a love for 'simplicity' and simple ideas: if the cause of social evil is 'impurity,' then the solution is the removal of the offending substance or subject.

Romantic ecologists also have the tautological argument of 'natural law' on their side. If nature is pure, then it is lawful and 'natural' that such purity shall pervade. Why should there be population control? To protect the natural limit of resources of the planet. *It is only natural* that there should be so many people on the planet. Ecology is the perfect environment for the cultivation of a purist critique of 'modernity'. Its green pastures provide free reign for the unbridled advance of a theory which provides both moral and scientistic ground for a critique of both modern and post-modern society.[11] Within the green expanses of ecology, the wild imagination of the nature romantic can run free with the certainty that what was old was not only good, but most importantly, it was 'natural.'

The longing for an ecologically pure society reflects the desire to return to a time and place when society was free from the decadence associated with urban life. There is a distinctly rural bias within ecological discourse, a depiction of the rural landscape as a vestige of past golden age of ecological purity and morality. Since the emergence of capitalism and the arrival of the urban capitalist center, the gap which opened between a world that had been largely agrarian and an increasingly urban society provided a space for the purist's romantic reverie. Often a bourgeois urbanite and rarely directly engaged in agricultural work, the nature romantic wrote about the abstract goodness of a rural life of the past, longing for an end to modernization and urbanization.

However, the story of the town and country divide is hardly one of good and evil: while the country has not always constituted a realm of innocence, the city has not always been such a bad thing. As Raymond Williams points out in the case of Britain, the real histories of the 'country way of life' and 'city life' are astonishingly varied and uneven.[12] While the rural village is often associated with ecological well-being and social cohesiveness, there exists a less liberatory association with the rural village that is not commonly discussed within contemporary ecological discussions.[13] The parochial tendency of rural life has often been a source of alienation for the stranger as well for those viewed as strange within the village itself. Women, gender-benders, those with a vision that extends beyond the scope of the close knit community, have often been suppressed by the homogenizing tendency of small village life. Standing in sharp contrast to the harmonious and wholesome portrayals of 'country life' are such parochial European rural disasters as the Spanish Inquisition, European witch burning, Eastern European Pogroms, and U.S. plantation slavery—atrocities that often took place within pastoral, 'natural' rural contexts.

In turn, while much contemporary ecological discussion portrays the city as a center of industry, pollution, and social alienation, it has also represented a haven of social freedom. Out of the broken ties to family and village, came as well the opportunity to encounter new ideas and liberties. It is within cities that many social movements have emerged over the centuries, providing a refuge for those who were not always accepted within parochial rural villages such as Jews, Gypsies, intellectuals, secularists, anarchists, artists, and sexual non-conformists. While rural life undeniably offers the potential for close community ties and a closer tie to the land, it can also prove hospitable to xenophobia, social conformity, and parochialism.

Despite the heterogeneity of categories of 'city' and 'country', there still exists a strong rural bias within ecological discourse. For example, a generic description of 'ecotopia' is primarily located within a rural environment. The inhabitants of that imagined ecotopia are usually wholesome, able-bodied, white, and heterosexual. These taken-for-granted associations latent within popular consciousness are often shared particularly by European descendants raised within industrialized capitalist societies that define 'nature' in opposition to society and the evil town in opposition to the wholesome country. Rarely would one imagine the 'ecological subject' to be a Puerto Rican lesbian in the Lower East Side of Manhattan, a poor disabled man of color in Chicago, or a Jew in Brooklyn, for ecology is primarily defined in opposition to the urban subject. The predominantly urban identity of such progressive movements such as feminism, lesbian and gay liberation, civil rights, and labor movements, renders feminists, queers, Jews, people of color, and urban workers as incongruent with white middle-class 'wholesome' understandings of 'ecology'.

Implicit within the rural bias which marks much ecological discussion, is a reactionary nostalgia for the goodness of 'the simple life' of the past. Today, the old guy on the Quaker's Oatmeal commercial suggests that living simply is "the right thing to do." An Emersonian nature romanticism wafts through the air, informing us that all we need is a simple house, a good book, and a chestnut or two to roast on the fire. It is time, we are told, to end our years of debauchery, time to buckle down. The family is re-romanticized as in the fifties, babies are 'in' and 'family values' must be restored.

This romantic rurally biased 'conservationism' smacks of political conservatism. A recent ad put out by Geo says, "In the future, more people will lead simpler lives, protect the environment, rediscover romance and…get to know Geo." The full-page ad presents a black-and-white photograph of a home-town looking teenage boy and girl relaxing wholesomely in a convertible. The girl sports a fountain of long blond flowing hair, her face clear of make-up, and reclines with the boy, wearing clothing lifted directly from the late fifties; a time when the country was still 'innocent'. The ad suggests that it would be desirable to restore the simplicity of the days before the Vietnam War, the civil rights and women's movements. 'Romance', which the women's movement is blamed for destroying by challenging gender roles, will be restored as well. Environmental campaigns increasingly conflate the decadence of today's neo liberal capitalism with yesterday's New Left, citing the latter as the cause of social and ecological breakdown.

However, there is nothing romantic about living simply. Women and the poor have lived the real 'simple life' for centuries, impoverished by economic and social institutions of compulsory heterosexuality and alienated labor. A life without choices, alternatives, and in many cases, material subsistence, is indeed very simple. Our world is becoming increasingly culturally impoverished and simplified, filled with senseless commodities and spectacles. Women and all marginalized peoples, at the center of this quality crisis, cannot afford to live any more simply. And because so many have lived simply, restrained by authorities for centuries, the romantic appeal to conserve nature sounds seductively familiar; so familiar that many accept such admonitions without even thinking. However, upon closer look, we see that we are being implored not to release human potential for social and political transformation within society but instead, to 'conserve' nature.

Consumer Ecology: The Romance Of Ecological Self-Constraint

The desire for a pure, 'simple' social world has claimed a new theater within contemporary society, this time wearing the mask of the ecological consumer. Within this contemporary play, the well-meaning purist yearns to slay a new dragon: the impure product. For those who feel demoralized and poisoned by social and ecological degradation, consumer ecology offers a way to combat the dragon of ecocide'while purifying the body and soul at the same time, all without destablizing institutions such as the state, capitalism, or racism.

The search for an ecological life style reflects the longing to establish congruence between consumption practices of everyday life and ecological ideals. Consumer ecology expresses a scientistic dimension of ecology, dictating methods of environmental and physical 'hygiene' loaded with moral and spiritual meaning. Practices such as recycling, energy conservation, veganism, vegetarianism, or consuming organic products, are considered not only physically and environmentally more healthful, but resonate with the moral desires to be pure of spirit as well.

Consumer ecology is a discreet 'private practice' articulated within the dialogue between private industry and the private domestic sphere: a private response to the popular observation that both these spheres have been degraded and must be purged.

Consumer ecology is a postmodern brand of asceticism based on romantic values of idealization, protection and constraint. Promoting an *idealized* commodity that is chemical and waste-free, consumer ecology encourages the never ending search for the 'pure commodity' that contains as much 'pure nature' as possible, while making the least impact on the natural world.

In turn, the preoccupation with *protection* is deeply embedded in the world of commodity purity as well. Eco-consumers and green capitalists alike express their value of *self-constraint* by exercising self-control in the production and consumption of impure commodities. Upholding this impulse is the belief that down deep we are all greedy consumers who must restrain the desire to over-consume. Just as the courtly troubadour demonstrates desire for his lady by promising sexual self-constraint, individuals in society are encouraged to express their desire for nature by promising to constrain their inclination to spoil and deplete the environment.

The impulse toward romantic self-constraint assumes a variety of forms, ranging from self-restraint regarding consumption to reproductive restraint. At the more benign end of the spectrum, corporations appeal to individuals to restrain their everyday appetites for 'natural resources'. Advertisers often deploy emotionally laden images of nature in their attempt to evoke in individuals a sense of shame and accountability for the destruction of the natural world. For example, a few years ago, a TV campaign by Pepsi depicted a sentimental image of baby ducks swimming in a reedy pond with small children playing in the sand nearby. The caption read in pink script, "Preserve It: They Deserve It." Through the use of soft lenses and young children, Pepsi effectively associated the idea of nature preservation with an underlying injunction against defiling innocent children. The Environmental Defense Fund had a recent TV commercial in which the camera zoomed in upon the hands of a white man crumpling a 'whole earth' photograph. As the earth's image was reduced to a tight paper ball, a stern voice announced dryly, "If you don't recycle, you're throwing it all away." In both instances, the message was clear: If individuals fail to constrain their desire to 'trash' nature, the natural world is done for.

Green capital participates in the cult of romantic consumption, promoting collective self-constraint on the part of consumers. Stonyfield Farm for instance, recently launched a campaign called "Planet Protectors" which makes a romantic plea to children to change their own unchivalrous ways as well as those their parents. Planet Protector's mascot is a cartoon cow soaring through the air like superman, cape and all, ready to save planet earth. The theme is clear: by re-using Stonyfield Farm's plastic yogurt containers, we all can protect the planet from harm. In their quarterly "moosletter" they ask their young readers: "Are you a planet protector? Are you committed to taking ACTION to protect and restore the Earth? Do you act in ways that protect Earth from harm and heal damage already done?"[14]

After providing information regarding the status of tropical rain forests (whose living things, they report, include only plants and animals, no mention of people), they explain "tropical rain forests are rapidly disappearing due to logging and other development." As for the solutions to these problems, Stonyfield Farm encourages children to "make a difference" by choosing to "use public transportation, carpool, walk, and don't leave lights on when you're not using them." Finally, the children are warned "every time you flick on a light or go for a ride in the car, $CO2$ is released into the atmosphere from the coal, oil, or gas burned to make energy. Be a planet Protector!"[15]

On the surface, Stonyfield's message seems reasonable enough: we should each do our part to save the planet. However, it is what is left out of the message that is deeply troubling. First, by failing to discuss the human suffering of peoples living within the 'natures' they represent, they separate the ecological from the social, blaming the entire society for ecological harm. Second, Stonyfield individualizes the problem by making no mention of institutional causes of ecological degradation such as capitalism, government, the World Trade Organization, or the military industrial complex (responsible for an overwhelming majority of pollution and resource extraction). Children are led to believe that by failing to restrain their individual hungers for car travel and electricity, they are as responsible for causing and solving ecological problems as are those unidentified institutions responsible for logging and other development.

In the more extreme wing of the ecology movement, individuals are warned to restrain not only consumption practices, but sexual reproduction practices as well. In such discussions, the mere presence of 'humanity' itself (resulting from an 'unrestrained' fertility) is cited as the cause of ecological injustice. According to the "Voluntary Human Extinction Movement" (VHEM), individuals should express a love of nature by endorsing voluntary childlessness. On their home-page on the Web, the VHEM presents a series of brief question and answers about the movement presented in a light and jocular style that explains their philosophy. According to "Les U. Knight," the movements' "spokes organism," the human "experiment" has run its course:

> The hopeful alternative to the extinction of millions, probably billions, of species of plants and animals is the voluntary extinction of one species: Homo Sapiens...us. Each time another one of us decides

to not add another one of us to the burgeoning billions already squatting on this ravaged planet, another ray of hope shines through the gloom. When every human chooses to stop breeding, Earth will be allowed to return to its former glory, and all creatures will be free to live, die, evolve (if they believe in evolution), and will perhaps pass away, as so many of Mother Nature's "experiments" have done throughout the eons. Good health will be restored to Earth's ecology...to the life form known as Gaia. It's going to take all of us going.[16]

According to the saddening reasoning of VHEM, 'humans' are so flawed as a species, so inherently carnivorous and unrestrainable, they will inevitably devour the planet. The only way to address this irrestrainable nature is for an ambiguous 'us' to phase out 'humanity'.

At the even more extreme end of the movement stand blatantly reactionary groups that advocate authoritarian measures to eradicate 'humanity' itself. The Gaia Liberation Front (GLF) asserts that "all life on planet Earth is more important than the survival of the human race."[17] According to their 1997 mission statement, "the total liberation of Earth can only be accomplished through the extinction of the Humans as a species."[18] Yet unlike the VHEM, the GLF endorses "involuntary" genocidal tactics including involuntary mass sterilization as well as the release of "anti-Human" viruses such as the airborne version of AIDS. According to "spokes organism" "Geophilus" whose writings can be found in their Web home page, authoritarian tactics are the only option for restoring ecological integrity:

The evidence is overwhelming that the Humans are programmed to kill the Earth. This programming is not only cultural, but probably also genetic since the major technologies Humans use for this purpose, from agriculture and metallurgy to writing and mathematics, have all been invented independently more than once. In any case, Human now carries the seeds of terracide. If any Humans survive, they may start the whole thing over again. Our policy is to take no chances.[19]

What makes this expression of ecology particularly troubling is its appeal to the concept of an innately flawed 'human nature' that must be cast *in toto* out of the 'garden'. Unlike other reactionary tendencies which blame particular social groups or 'technology' for ecological injustice, longing for a pre-fall industrial era, this group sees no possible 'return' or salvation for any sector within humanity. Invoking scientistic language deployed by Nazis (terms which describe humans as "vermin," or as "an alien species genetically programmed to kill Earth"), the GLF attempts to legitimize its claims by assuming the authoritative voice of the human technocrats they so condemn.

Of course most ecologically minded peoples do not present such extreme dictums for self constraint. Pleas for total reproductive restraint stands in sharp contrast to Stonyfield's reasonable request for individuals to turn off lights when leaving a room. Yet a common theme pervades the thinking of such romantics for whom true love can only be demonstrated by constraining the desire to defile nature. According to the romantic, the betrayal of nature results from a refusal of individuals to restrain themselves by failing to curb the tendency to consume, reproduce, pollute, and waste inherently scarce 'resources'. However, we must ask ourselves, is environmental degradation a mere betrayal of nature caused by the failure of individual self-constraint? Or is this degradation caused by a system of social institutions which allow a privileged few to denigrate and betray most of humanity and the rest of the natural world?

The environmental call for individual self-constraint implies a pessimistic view of society's potential relationship with nature. It suggests that our relationship with the natural world is inherently predicated on a repression of an inherent desire to destroy, rather than to enhance, natural processes. The idea of love as self-constraint reduces the idea of love to a holding back, or to a repression of a destructive desire rather than as an articulation of a social desire to participate creatively in natural and social processes. Thus we fail to see that we can actually cultivate new desires to create a just society where there would be neither helpless 'ladies' nor helpless 'mother natures' to protect. Privileging the idea of self-constraint obscures the idea of society's potential for rational ecological self-expression necessary for creating a world free of social and ecological denigration.

Romantic Concealment: Revealing The Nothingness Of The Banana

While allowing people to lighten their anxiety about ecological problems, consumer ecology is predicated on romantic concealment. Just as the knight's idealization of his lady conceals his underlying desire to maintain his own social privilege, the idealization of pure commodities conceals consumers' (often unconscious) desires to maintain their own privilege within a global capitalist economy. The mythology of a pure commodity based on consumer and producer protection and constraint conceals the deeper reality of a grotesquely immoral economic system which is sucking the very life out of the planet, along with over ninety percent of its inhabitants. Puritanical consumers who can afford to buy costly 'ecologically friendly' commodities can retreat into the discrete world of consumer heaven, where they are absolved of the sin of impure consumption. Focusing on the content of consumption allows consumers to remain within the kingdom of consumer heaven without looking down to see the very hell that capitalist production makes of the earth.

Carol Adams explores a similar problem of 'concealment' in her book, *The Sexual Politics of Meat: A Feminist, Vegetarian Critical Theory*.[20] In this work, Adams describes the concealment of the grim realities of the meat industry within capitalist patriarchy. Adams describes this concealment as the fabricated nothingness of meat, a popular perception shared by most consumers of factory-farmed meat products. According to Adams, vital to an ecological ethics is a challenge to the fabricated belief that meat is "nothing":

> ...awareness of the constructed nothingness of meat arises because one sees that it came from something, or rather someone, and it has been made into a no-thing, no-body...In experiencing the nothingness of meat, one realizes that one is not eating food but dead bodies.[21]

Adams calls feminists and all meat eaters to challenge the idea that meat is 'nothing', to reveal the cruelty and immorality of factory farming and of meat-eating in general.

As we deepen our social analysis of production practices in general, we see that the idea of the "nothingness of meat" may be extended to reveal the "nothingness of commodities" in general. Just as meat-eaters often fail to appreciate the subjectivity of animals that are plundered by factory farming, consumers in general fail to recognize the subjectivity of the people who are exploited in the production of commodities in general. For instance, while people are often unaware of the suffering of the factory farmed calf when they buy a plastic-covered slab of veal, they are often unaware of the struggle of women workers in a multi-national textile industry that produce the very shirts on their backs.

In addition, when we consider the social and ecological devastation caused by agribusiness, we see that the consumption of vegetable products is often as immoral as the consumption of animal products. For instance, a banana is not always a more moral food choice than a chicken. If we look at the social and economic relationships that transform bananas and chickens into commodities, we often uncover a far more complex set of social problems which determine whether the chicken or the banana represents a more 'moral' food choice. When we reveal the social context of banana production, we are confronted by a moral paradox: while the *content* of the banana (a form of non-sentient plant life) may represent a moral food choice, the *social relations* surrounding the agricultural production of a factory-farmed banana, may render such a food choice immoral.

When we reveal the nothingness of a banana, we become aware of the truly lethal social and ecological realities that deliver the banana from the Third World to the First. Most bananas sold in the First World constitute a cash crop which many Third World countries export in order to repay their debt to the World Bank or to the International Monetary Fund. These crops are cultivated on soil which could be used for the cultivation of foods for the local community itself. Consequently, people across the Third World literally starve while their land is controlled and converted to export zones for cash crops such as fruits, vegetables, sugar, tobacco, coffee, and timber. Agricultural workers are paid slave wages, denied health benefits, and are exposed to pesticides, herbicides, and chemical fertilizers (bananas are one of the most pesticide-toxic fruits).[22] Certainly the agricultural worker, who is poisoned with over-work and chemical imputs, whose indigenous land was first confiscated by colonialists, then repossessed by the World Bank, should be given the moral consideration that many vegetarians would give to the chicken. Yet it is often easier to reveal the 'nothingness of meat' than to reveal the 'nothingness of workers' or the 'nothingness of cultures' that are degraded by producing bananas.

As we recognize the complex and contradictory nature of capitalist production, it becomes clear why activism regarding the unethical consumption of meat often exceeds activism regarding the unethical consumption of commodities in general. While animals have been reduced to a specific commodity that we may eliminate from our diet, commodities in general thoroughly permeate our social world. It would be impossible to expel each one from our daily lives. The fact is, within a global capitalist system, we are largely unable to determine the modes and ethics of production. It is understandable, then, that many of us focus on areas of consumption (such as diet) over which we feel we can exercise some control. However, the longer we focus on the ethics of consumption, *as if we could consume morally within a capitalist system*, the longer before we reveal the inherent immorality of the capitalist system itself.

The desire for 'nature', the desire for ethical organic practices such as food production, must be broadened and deepened to include as well, a desire for social and political freedom. The desire to spare animals from disrespectful and harmful practices must be elaborated to include an overall challenge to a capitalist system that threatens the very survival of people. Once we reveal the 'nothingness' of the commodity, overcoming what Marx called "commodity fetishism," we will recognize that each commodity, as Adams says, "came from something, or rather someone, and it has been made into a no-thing, no-body."[23] In recognizing the fabricated nothingness of the commodity, we realize that we are not merely consuming abstract commodities but that we are devastating actual people's lives, land, and cultures. Ultimately, it becomes immoral to separate contents of consumption from forms of production; for in so doing, we turn our heads from the social, ecological, and political costs of global capitalism itself.

The Romance of Techno-Dragons: The Fight To Slay 'Technology'

Accompanying the struggle for 'pure' commodities, has emerged the struggle for pure technologies. Despondent about the degradation of ecological and social life, people look to the most obvious visible tropes of modern and postmodern society: technology itself. Noting the historical correlation between 'advanced' technologies and the reduction in quality of life, people create causal connections between 'technology' as a general category and ecological injustice in particular. In search of solutions, many look longingly to a past golden age where 'low' technologies did not plunder the earth's riches; a time before the dragon of 'modern technology' bore its mechanized and treacherous claws, destroying all that it encountered.

Yet today's romantic discussions concerning modern 'technology' really reflect crises concerning capitalism and democracy: crises in which citizens are deprived of political forums in which to shape the forms and functions of capital driven technologies. All around us, we see new technologies sprout up within *Newsweek* or on the nightly news. Yet we play no direct *political* role in determining what effect they shall have upon our social and ecological lives. The technologies which most concern us tend to be referred to as 'high' or 'industrial' technologies, technologies whose deployment requires intensive degrees of centralized capital or labor, often at the expense of both social and ecological integrity. Hence, computer, nuclear, communications and biotechnologies, represent sources of tremendous concern for those concerned with social and ecological justice. However, when we remove such discussions from their calls to 'go back' to earlier, easier times and places, we see a different set of problems and opportunities emerge. By exploring the social and political context of these 'high' technologies, we see that they are after all, capitalist commodities produced by corporations, regulated by the state, and often originally researched and developed by the military.

So often, 'backward-looking' discussions portray 'technology' as a universal event that emerges within a social and political vacuum. We live in an era of technological determinism in which we are told that 'technology' exists as an autonomous force which *determines* social and political events. Today, we become familiar with ideas of technical determinism in journalistic stories which speak of "technology out of control," or "computers transforming the world" exemplified by the opening of this *Newsweek* article:

The (computer) revolution has only just begun, but already it's starting to overwhelm us. It's outstripping our capacity to cope, antiquating our laws, transforming our mores, reshuffling our economy, reordering our priorities, redefining our workplaces and making us sit for long periods in front of computer screens....Everything from media to medicine, from data to dating has been radically transformed by a tool invented barely 50 years ago. It's the Big Bang of our time.[24]

Such narratives present the idea of 'technology' as a self-driven force within 'humanity' which can shape or level a social world with the same power as a giant meteor. For the technological determinist, it is not economic or political institutions which reshape our practices of media, medicine, economy, law, and morality: It is the autonomous and unstoppable 'advance' of 'technology' which demands that we either get 'wired' or get wasted.

By regarding technology as a general 'human' force or a universal dragon, we fail to locate specific institutions which design, finance, and deploy harmful technological practices. Too often, no one is to blame when a technology goes wrong. Instead, each ecological disaster is portrayed as a case of technology out of control. Or, worse, when we do identify individuals or institutions as accountable for disaster, our analysis often remains too narrow: when the Exxon Valdez spilled its lethal tons of oil, the drunk driver of the oil rig was identified as the guilty party rather than the broader institutions of capital and state apparatuses which stress and regulate workers and natural processes for profit. When we blame technology in general, not only do we fail to identify corporations who financed the technology, but we fail to identify the state who granted the patent, and subsidized the corporation, excluding citizens from the decision making process.

The truth is, talking about technology is often an excuse for not talking about institutionalized power. It is often an excuse for not talking about the specific ways that institutions such as corporations and the state collude in shaping technologies that are socially and ecologically unjust. It is an excuse for not talking about the lack of real democracy. And what do we gain by talking about 'technology' instead of talking about capitalism and the state? We comfort ourselves with the romantic illusion of being institutionally oppositional when in fact, we actually *support* capitalism by providing new opportunities for corporations to *diversify* their markets by creating 'soft', 'low impact', and 'environmental friendly' technological alternatives for the rich which exist alongside of the really dangerous ones.

We cannot fight social institutions merely by critiquing *social mediums*, or the material expressions of culture. Just as art and language represent social mediums, technology is a social medium that represents a cultural practice of *technics* or a prosthetic engagement with the world. Social mediums such as art, language, and technology are often determined by social institutions such as the state, capitalism, or patriarchy. For example, today, while corporations, the state, and universities determine much of what will be considered 'high' art, they also determine what will be considered 'high' technology. Although there exist popular grassroots artists and technicians who maintain degrees of autonomy from large hierarchical institutions, their cultural practices impact far less dramatically upon society than those subsidized by powerful institutions. In France, language is actually controlled by the patriarchal state which manages and sustains not only highly gendered linguistic standards, but the incorporation of foreign language and food as well.

However, while it is wrong for the state, corporations, or universities to autocratically determine *any* aspect of social media, we cannot abolish authoritarian institutions merely by protesting against language, art, or technology per se. Attempts to *upgrade* social media by creating for instance 'a feminine language', 'a people's art', or a 'low technology', fail to eradicate the source of control of social media. Whereas we may create the alternative of a feminine language, there will still exist patriarchy and the state which oppress women. Similarly, while we may create a people's art or a low technology, we will still be confronted by a state, a corporate edifice, and an educational system which controls our lives and destroys the earth in a vast array of other dangerous ways. Finally, proposing 'low' technologies, while opening up potentially thoughtful dialogue regarding the ethics of technology, does little to oblige people to consider the political and economic conditions which allow corporations and governments to autocratically create social and ecological injustice in the first place.

What is more, the 'lowness' of a technology does not determine the justness of its social application. Despite romantic dreams of the inherent goodness of technologies of the past, there exists much in our technological history that is to be desired. As Bookchin points out, while the pyramids in Egypt were built by slaves using very low technologies, early American settlers clear-cut miles of native forest merely by burning and felling, as opposed to using the "high-tech" chain saws of today.[25] Furthermore, before implementing the 'higher' and more efficient modern technologies of mega gas-chambers, Hitler was quite effective in using simple bread trucks and exhaust hoses to round up and asphyxiate entire villages of Jews (before 'advancing' to gas chambers). Clearly, we could not say that technological 'advance' was the determining factor for the death of six million Jews. Rather, it was a set of *social relationships* that allowed for the horrific collusion between a fascist state, a racist ideology, a legacy of anti-Semitism, and an entrepreneurial factor, giving way to genocidal devastation. We must consider the absurdity of fighters in the Polish resistance protesting the Holocaust on the basis of objections to the high 'technology' of gas chambers alone.

Low technologies that are supposedly fulfilling a benign function, are not always liberatory on a social level. Along the coast of Northern California, stretch miles of gargantuan windmills: while representing a 'low' technology, these monstrosities also represent the state's techno-fix to the problem of doling out 'energy' in a centralized and bureaucratic fashion, blotting out the glittering sea shore along the way. Similarly, the enormous solar collectors in the Southwest represent a low technology of preposterous proportion. Rather than promote local and direct expression of technological ethics, such large scale technologies promote instead the centralized power of the state and corporations who engineer and execute the design of their own choosing. It is indeed crucial that our technological practices do not degrade natural processes. Yet it is also necessary that we do not harm the social world by usurping community self-determination. There is no recipe for a 'good' or 'ecological' technology independent of a truly democratic context.

So, we might ask, if technology is not deterministic, if it is informed by particular social relationships, is it in fact simply 'neutral'? Are technologies blank slates to be written upon by those in power? Nothing could be farther from the truth. While there are many technologies, such as a knife, which contain a wide spectrum of potential functions, good and bad, there are many technologies which by their very design are 'loaded' in positive or dangerous ways.[26] For instance, a nuclear bomb is structurally biased by its design and function to kill inordinate amounts of people quickly or to 'peacefully' intimidate political leaders into submission. However, while we might say that a nuclear bomb is not neutral we could not say that the technology of nuclear bombs alone *determined* the events in Hiroshima or Nagasaki.

Although the nuclear bomb represented a necessary condition for the nuclear bombing of Japan, it did not constitute a sufficient condition. The sufficient condition was comprised of a set of social relationships: a hideous amalgam of foreign policy and a technological expression of that highly undemocratic and capital driven system, called 'nuclear technology'. Given enough time, money, and undemocratic power to develop 'technology', those in authority can dream up some pretty lethal inventions.

Similarly, organic fertilizer is structurally biased in a clear direction, albeit a positive one. It is constituted by the very intention underlying its design to enhance, rather than deplete, the composition of soil and water. However, while we might say that the technology of organic fertilizer is not 'neutral', we could not say that the technology of organic fertilizer will actually determine that the world's soil and water will be enhanced. Rather, it is a set of social relationships that determines the scale by which agricultural workers will be able to apply organic fertilizer, as well as whether the soil and water will be too damaged by previous chemical abuse. Hence, whereas organic fertilizer represents a necessary condition for an ethical and ecological agriculture, it alone represents an insufficient condition. The sufficient condition for a liberatory organic agriculture is a social and politically just context: the reconstruction of political and social institutions which not only ecologize, but democratize agricultural practice.

The Techno-Fix: Slaying The Techno-Dragon

At this juncture we might ask ourselves: why are there so few discussions which explore questions of institutional power in regards to technology in the Ecology Movement? Why have ecological discussions of technology tended toward romantic dreams of slaying dragons of 'modern technology'? Why would so many in the ecology movement prefer to critique the universal category of 'technology' in general as a social medium, rather than critique the political and economic social relations which engender particular technological practices?[27]

Many of us who grew up in post-cold war America have little consciousness of a revolutionary tradition. Few are aware that there existed a time before the state or capitalism. We accept these hegemonic institutions as inevitable, irreplaceable, and taken-for-granted. Therefore, when we are moved to critique society, we focus on questions of social mediums we believe we can change, rather than on social or political relationships and institutions which we see as universal and insurmountable.

Romantic yearnings for 'low' technologies tend to lead to some pretty ironic outcomes. A few years back, neo-Luddite Kirkpatrick Sale enacted his anger at 'technology' by smashing a computer on stage at New York's Town Hall. Now surely, Sale knows as he takes a hammer to the machine that the computer possesses no autonomous institutional social power. He knows that the computer is neither neutral nor technologically determined, but that it represents a social medium, a social-technological *expression* of the institutions of the military, the state, and corporations such as IBM or Microsoft. By smashing the computer in the social forum of New York City's Town Hall in Manhattan, Sale tells us that he wishes his critique to be social if not explicitly public. Yet Sale belongs to no municipal political forum in which his position regarding the goodness or badness of computer technology has any authentic political power. Rather than express his voice politically, Sale's voice is rendered *spectacular* as the glossy (computer enhanced) photograph of him heroically slaying the computer on a page of *Wired* Magazine (a computer users' publication).

If Sale were to think socially and politically, rather than romantically, about the computer he smashes, he might think about how, while it might feel cathartic to smash the computer, there might be still more oppositional ways in which to express his sentiments regarding computer technology.[28] Rather than smash the computer with a sledge hammer, were Sale to critique the lack of economic democracy surrounding the computer industry, he might have considered the fact that only privileged people gain access to computers, such as those working at the press which publishes his books. Instead, Sale might have thought to perhaps share his computer, for instance, with a community center some forty blocks down in the Lower East Side, called *Charas*, where radical activists in the Puerto Rican community are engaged in oppositional work for social, ecological, and political change. Activists at a non-profit organization like *Charas*, who may not be able to afford a costly computer, might be able to use the machine to publish a newsletter for the activist community or might use it for some other activist project.

After giving his computer to activists at *Charas*, Sale could have then joined his neighborhood association where he could have engaged in a political debate regarding the social and ecological ethics of computerization while discussing too, the need for direct democracy. He could have discussed the need for political forums in which we all may participate in making decisions regarding an even broader spectrum of social and technological issues. Rather than point his weapon at the dragon of technology, industrial society, or mass society, he could have discussed how computer technology is driven by an undemocratic global *capitalist* economy. Moreover, he could have assisted others in understanding how capitalism in general dehumanizes people and destroys the rest of the natural world. In short, if Kirkpatrick Sale were to talk about social relationships rather than generalized social media such as 'technology', he would talk about computers in the context of such institutions as the state, capitalism, racism, and sexism. However, were he to take such a position, would he have ended up being featured in *Wired* magazine?

Each of us must ask ourselves such difficult questions as we enter discussions concerning technology, or any social medium, for that matter. We need to constantly ask ourselves: are there necessary pieces of the picture that we leave out, and why? The fact is, we can often glean more support for critiquing a social medium such as technology (or for slaying vaporous dragons such as mass society or industrialism) than for attempting to abolish and transcend social institutions such as the state or capitalism. We must extend our critique beyond social mediums because social institutions exist *prior* to and *independent* of such mediums. For example, while merchant and rural factory capitalism emerged as a dehumanizing system prior to the emergence of industrial capitalism, the state preceded the emergence of capitalism itself. The desire to eliminate 'high' technology therefore, is not just insufficient for creating a free and ecological society; it also shifts the focus from the real problem of undemocratic, dehumanizing, and anti-ecological social institutions.

And so the question remains: just because we have no direct democratic control over our economies or state (and thus over technological practice), do we cease to critique technologies which we esteem to be socially and ecologically dangerous? Are we obliged to choose between a critique of technology *per se* and a critique of the state or capitalism? Clearly, the answer to these questions is no on both counts. Questions concerning technology may allow us to broaden our thinking about the lack of political and economic democracy surrounding particular technological practices. We can explore the *specific* harms of particular technologies, calling for social and political action, while broadening our understanding of the political and economic context in which we have little control over capitalist and state practice. In this way, each *specific* issue concerning technology provides a forum to speak *generally* about the need for economic and political democracy. Each time we talk about a specific technology or about technology in general, without discussing the urgent need for political democracy, we miss a vital opportunity to raise consciousness regarding the broader context of social or ecological change.

For The Love Of Nature: Knowing Self, Knowing Other

In love, there is a paradox. In order to know and understand that which we love, we must first know ourselves. We must engage in a continual process of becoming conscious of our own beliefs, prejudices, and desires if we are to truly see that which we love. When we fail to know ourselves in this way, the beloved can be nothing more than a projection of our own desires, a projection that obstructs our vision of the desires, history, and distinctiveness of those we love.

In order to truly love nature, society must know itself; it must understand its own social, political, and economic structure, understanding in turn how each individual benefits or suffers from such structures.

Yet instead of knowing society, many in the ecology movement tend to focus exclusively on an idea of 'nature' that has become the small blue pool into which Narcissus gazed, enamored by his own reflection. Rapt with his own image, Narcissus saw neither the color of the water, nor did he feel its coolness against his fingers. In the same way, when the privileged look into the 'pool of nature', they too, cannot see what grows there. They cannot see 'nature' as a contested political and social ground whose abundance and scarcity are unevenly distributed. Instead they see only the romantic reflection of their desire to preserve the institutions and ideologies that grant them access to both social and ecological privileges; they see only the image of 'mother earth' as a nurturing victim in need of their protection and control.

The practice of authentically 'knowing' nature is one of politicized critical self-consciousness. As social creatures, we look at the world through social eyes. In order to see nature, we must be increasingly conscious of how our understandings of 'nature' are shaped by historical institutions such as Christianity, capitalism, racism, and patriarchy which give rise to contradictory yet persistent notions of nature as pure, greedy, competitive, dark, passive, and nurturing. For instance, if we are not conscious of the social-religious causes of our own social guilt and self-hatred, we will romanticize nature as a pure and superior being before which we feel puny, humbled, and wretched. In the same way, if we do not transcend "internalized capitalism," a hegemonic acceptance of capitalism as normative, inevitable, and progressive, we will continue to portray nature as a social Darwinian nightmare: a romantic drama in which only the strongest knights, or those best able to make a buck, can survive. In this shameful narrative, the privileged turn their backs on the 'poor majority' who carry both the brunt of and the blame for ecological injustices. In contrast, a radical love of nature entails that we become aware of the history of ideas of nature in addition to politically resisting social hierarchies that nurture distorted understandings and practices of nature as well.

In particular, we must extend this critical self-consciousness to our poetic and visual expressions of our desire for nature. We must be critical of our use of metaphors and images of natural processes, making sure that they do not reproduce racist or sexist cultural stereotypes. While there are indigenous cultures that appeal to non-sexist female images of nature, when members of non-indigenous cultures attempt to deploy 'mother-earth' metaphors, something vital is lost in the translation. Indeed, a metaphor which emerges within the language of an indigenous people cannot always be translated into the language of a culture that emerged in an era of modern and postmodern capitalism.

Audre Lorde points to a similar linguistic difficulty when discussing the slave who uses the "master's tools" to dismantle the master's house.[29] This has been an ongoing struggle especially for ecofeminists relying upon patriarchal language and philosophical constructs to critique and reconstruct patriarchal discourses that relate to ecology. Often, the origin of words and their historical relationship to oppressive ideologies actually contradicts the very spirit of liberation that ecofeminists attempt to convey. Within the current society, female metaphors of nature cannot be abstracted from Western patriarchal values, desires, and definitions of women that saturate media, religion, and educational forums. The metaphor of 'mother-nature' is culturally loaded with masculinist ideologies that 'justify' women's compulsory heterosexuality, motherhood, and subjugation: It contains the history of what it has meant to be both a woman and a mother within this society.

Because we are social creatures, our understandings of nature will never be pure or free of social meaning or contingencies. Nature is not a thing from which we can separate ourselves and know completely, no matter how liberatory our culture or language may be. Instead of trying to grasp a romantic knowledge of a people-less 'nature' through abstract love, protection, and contemplation, we must begin to know and reconstruct the social and political institutions that determine both social and ecological practices. By engaging in a life long process of politicized critical self-reflection and action, we may become a society conscious of the historical origins of its own desire for 'nature'; a socialized desire that begs to be developed in a truly radical direction.

Notes

1. Denis de Rougemont, *Love in the Western World* (Princeton: Princeton University Press, 1983), pp. 106-7.

2. Roger Sherman Loomis, Trans. *The Romance of Tristan and Ysolt by Thomas of Britain.* (New York: Boyer Books, 1931).

3. *Ibid.*, p. 64.

4. For a discussion of the relationship between sabotage and agency, see Sarah Lucia Hoagland, *Lesbian Ethics: Toward New Value* (Palo Alto: Institute for Lesbian Studies, 1988). pp. 46-49.

5. Murray Bookchin, personal communication, 18 July 1984.

6. There have been a number of truly intelligent discussions of reproduction issues by feminists such as Betsy Hartman that address social and political considerations. See Betsy Hartman, *Reproductive Rights and Wrongs: The Global Politics of Population Control and Reproductive Choice* (New York: Harper and Row, 1987).

7. World Bank. 1993. World Development Report 1993. New York: Oxford University Press.

8. Murray Bookchin, "The Power to Create, the Power to Destroy" in *Toward an Ecological Society* (Montreal: Black Rose Books, 1980), p. 37.

9. Bill Devall and George Sessions, "Why Wilderness in the Nuclear Age?" in *Deep Ecology: Living As If Nature Mattered* (Salt Lake City: Peregrine Smith Books, 1985), p. 127.

10. For a good discussion of structural adjustment programs, see Bruce Rich, *Mortgaging the Earth: The World Bank, Environment, Impoverishment, and the Crisis of Development.* (Boston: Beacon Press, 1994).

11. For an in-depth discussion of the historical relationship between ecological discourse and reactionary thinking, particularly within the German context, see Janet Biehl and Peter Staudenmaier, *Ecofascism.* (London: AK Press. 1995).

12. Raymond Williams, *The Country and the City* (New York: Oxford University Press), pp. 35-13. For more on the dialectics of town and country, see Murray Bookchin, *Urbanization Without Cities: The Rise and Decline of Citizenship* (Montreal: Black Rose Books, 1992).

14. Earth First! Bumper sticker as advertised in their catalogue.

15. *Stonyfield Farm Planet Protectors Earth Action Moosletter.* Winter 1997.

16. The question of whether the Voluntary Human Extinction Movement is a satirical or sincere expression of anti-humanist views is debatable. The subtitle for their manifesto is "A Modest Proposal," a clear allusion to Swift's famous pamphlet which satirically proposed eating babies as a means of relieving Irish famine. However, whether they are exaggerating

Malthusian rhetoric as a means to expose its callous insanity, or whether they are sincere, the fact that so many take it seriously reflects a troubling state of affairs within the ecology movement.

17. Gaia Liberation Front. Web site: http://www:paranoia.com/coe/

18. *Ibid.*

19. *Ibid.*

20. Carol J. Adams, *The Sexual Politics of Meat* (New York: Continuum, 1991), p. 175.

21. *Ibid.*, p. 175.

22. For a closer look at issues of worker's health and safety related to Third World labor conditions, see *Women in Development: A Resource Guide for Organization and Action.* (Philadelphia: New Society Publishers, 1984). Also, For a broader discussion of the implications of Third World 'development' women's labor, see Gita Sen and Caren Grown, *Development, Crises, and Alternative Visions: Third World Women's Perspectives* (New York: Monthly Review Press, 1987).

23. Adams, *The Sexual Politics of Meat*, p. 175.

24. Steven Levy, "Technomania: The Hype and the Hope," in *Newsweek* 27 February 1995, p. 3.

25. Murray Bookchin. Lecture. Institute for Social Ecology. 11 July 1995.

26. Arturo Escobar. Lecture. University of Massachusetts. 8 March 1995.

27. For a wider discussion of the relationship between technology and democracy, see Richard E. Sclove, *Democracy and Technology* (New York: The Guilford Press, 1995). Although Sclove's book explores the democratization of technology *within* the context of a representative statist democracy, he does pose a series of crucial questions concerning the lack of technological democracy within the present context. Also see Bookchin's discussion of the social and political implications of technology in *Re-Enchanting Humanity* (London: Cassell. 1995). pp. 148-172.

28. Kevin Kelly, "Interview with the Luddite," in *Wired.* (3.06 June 1995., p. 166). In his *Wired* interview, Sale comments on the personal satisfaction he gleaned from smashing the computer:

> It was astonishing how good it made me feel! I cannot explain it to you. I was on stage of New York City's Town Hall with an audience of 1,500 people. I was behind a lectern, and in front of the lectern was this computer. And I gave a very short, minute-and-a-half description of what was wrong with the technosphere, how it was destroying the biosphere. And then I walked over and I got this very powerful sledge-hammer and smashed the screen with one blow and smashed the keyboard with another blow. It felt wonderful. The sound it made, the spewing of the undoubtedly poisonous insides into the spotlight, the dust that hung in the air... some in the audience applauded. I bowed and returned to my chair.

29. Audre Lorde, "The Masters Tool's Will Never Dismantle the Master's House," in *Sister Outsider* (New York: Crossing Press, 1984), pp. 110-113.

CHAPTER TWO

REFLECTIONS ON THE ECOFEMINIST DESIRE FOR NATURE

During the past several decades, strands of ecological theory have emerged reflecting diverse expressions of the desire for ecological integrity. By tracing the development of specific ecological discussions within a wider ecology movement, we may gain an appreciation for the challenges and possibilities that arise as particular groups begin to explore the connections between social and ecological justice.

As noted in the previous chapter, the desire for ecological integrity can be marked by moments of individualism, abstraction, and romanticism that can be traced back to ecology's European origins. Yet as this chapter illustrates, ecological activists may also express this desire in more social and political terms, linking problems of ecological degradation to questions of hierarchy and oppression within society. In such cases, the "desire for nature"—or the desire for a quality of everyday life that is healthful, meaningful, and ecological—is framed as a need to overcome social as well as ecological injustice.

Using ecofeminism as a case study, this chapter examines the process by which different groups approach ecological issues from a more social, rather than individualistic or romantic perspective, recasting questions of nature in terms that reflect their own identities and situations. It is through exploring the connections between ecology and social justice that ecofeminists ground their desire for ecological integrity in concrete social and ecological realities of everyday life. In so doing, ecofeminism is largely able to articulate a social desire for nature, transcending many of the limitations that mark the wider radical ecology movement as a whole.

Yet the history of ecofeminism has not been without hurdles. Emerging from a variety of different ecological and feminist tendencies, ecofeminists have often struggled, particularly in the early years, with questions such as how to avoid the tendency to invoke universal notions of gender, nature, and culture, or how to fit into a wider multicultural feminist movement.

This chapter explores a few of the primary trajectories by which ecofeminism originally unfolded in the 1980s. These "originating influences," radical feminism, social ecology, environmental justice and international environmental movements, reflect only several of the many movements that informed the development of contemporary ecofeminism. Yet by studying these tendencies, we may gain a general appreciation for the wider context in which women were beginning to approach the question of ecology in the 1980s, providing insight into the problems and possibilities that emerge as groups link questions of nature to issues of social, cultural, and political justice.

Radical Feminism And The Emergence Of The Body Politic

Within the radical feminism of the late sixties and early seventies, an organic sensibility began to germinate, eventually finding its expression within many ecofeminist writings today. This organic sensibility emerged within an exploration of the 'embodied personal' that found its first seeds within the context of the New Left.

Since the late 1960s, the body has become a touchstone to which many feminists return in order to measure the 'groundedness' of feminist theory. The body politic, developed by radical feminists, attempted to render feminist theory resonant with women's lived experience as flesh and blood in the world, providing a palpable praxis that corresponded with women's bodily reality. Ecological politics has also played a role in grounding feminist politics. Ecology, like the body, offers feminism an organic dimension by which to explore women's survival not as abstract 'sisters in patriarchy', but as women addressing the concrete and visceral dimensions of social and ecological injustice. And as we shall see, radical feminist body politics contains a latent ecological sensibility that, in turn, gives way to what would soon be called "ecofeminism."

In the late sixties and early seventies, thousands of women were involved in political organizations such as Students for a Democratic Society and the anti-war and civil rights movements. While participating in these struggles, many women brought to light glaring contradictions between the abstract principles and goals of political movements and their own personal, embodied experiences as women in the world. While men spoke of goals of liberty, freedom, and equality for 'humanity', movement women were often cloistered in the kitchen doing the mailings and making coffee for movement men. When women attempted to focus on their own liberation, they were often advised to wait for the 'greater liberation of humanity' at which time women's liberation would inevitably follow.

The women of the New Left soon grew tired of waiting. They began to recognize the contradictions between their own private, embodied struggles and the public, political ideals of larger struggles for social justice. Standing together in kitchens, or while licking envelopes, women began to engage in informal discussions regarding contradictions such as the irony of fighting against U.S. aggression in Vietnam during the day while often being abused physically at night by the same men who opposed the war. In a speech given at a city-wide meeting of radical women's groups in New York City in 1968, AnnE Koedt expressed women's dissatisfaction with leftist movements:

> Within the last year many radical women's groups have sprung up throughout the country. This was caused by the fact that movement women found themselves playing secondary roles on every level—be it in terms of leadership, or simply in terms of being listened to. They found themselves afraid to speak up because of self-doubts when in the presence of men. They ended up concentrating on food-making, typing, mimeographing, general assistance work, and serving as a sexual supply for their male comrades after hours. [1]

Women from all over the country formed groups where they could discuss their experiences in the movement and talk about the embodied details of their everyday lives. Some of these groups emerged into formal "consciousness raising groups" in which women began to see that insights and experiences once thought of as idiosyncratic or purely personal were shared by many others as well. Soon, like astronomers linking a seemingly random scattering of stars into a constellation, women began to link disparate personal experiences into a constellation of oppressions, which they referred to as 'patriarchy', that was highly political and historical in nature. Issues such as sexuality, relationships, health, work, family, and violence in the home and street, all once seen as women's personal 'bodily' issues not to be considered or discussed in public, now were examined and understood through a distinctly political lens.

Out of this analysis was born a "body politic," an attempt to understand the political implications of women's experience of male domination in their everyday lives. From this analysis came a radical feminist movement that created counter institutions to address the bodily dimension of women's oppression.

Women had begun to invoke new understandings of a 'biological' dimension of social life. All activities relegated to the domestic realm, the daily 'reproductive' biological activities such as cooking, cleaning, caring for the sick, bearing and nurturing children, and sexuality were now considered worthy of political attention. The great wall between the public and private realm shattered as women began to examine the organic dimension of their own work, lives, and ways of being in the world. In developing the dialectical body politic, women began to examine an organic dimension to social life unexplored by the wider New Left. It would not be long before the contradictions between the body and the rest of the natural world would be pressed to give way to an understanding of an ecological body that stands in direct relationship to a political, social world. Phrases including 'the personal is political,' 'sexual politics,' or 'body politics,' all reflected this new tendency to recognize the interconnections between the body and the political, shifting political discussion to include issues deemed 'organic' or 'embodied', reflecting an implicit ecological impulse.

To further contextualize this ecological impulse, it is crucial to locate radical feminism within the wider context of the New Left in which a new ecological movement was steadily emerging during the late 1960s. Indeed, during these years, an ecological sensibility had developed, reflecting a rejection of middle-class suburban values, aesthetics, and cultural practices. The publication of the *Whole Earth Catalogue* in 1968 heralded the arrival of a generation of youth seeking a new quality of everyday life deemed more organic, immediate, and "natural." The catalogue's pages offered "earthy" advice ranging from homesteading in the country to making natural soap in a spirit of ecology and "do it yourself" self-sufficiency. As a feminist correlate, *Our Bodies, Ourselves,* published in 1973 by the Boston Women's Health Collective, offered lay knowledge to women seeking self-sufficiency in the domain of reproductive health. The publication of both books signaled a time in which people sought asylum from a world they perceived as sterile, impersonal, and disempowering. The U.S. ecology movement spoke to these desires, providing "natural" alternatives for people striving to reconstitute a more healthful and self-determined quality of everyday life.

Along with this new ecological sensibility, there emerged within radical feminism an implicit anarchist sensibility as well: a critique of hierarchy in general that flowed from a specific critique of male domination. Seeking to incorporate this spirit of non-hierarchy into feminist projects and organizations, women adopted cooperative ways of working and relating together. By the beginning of the 1970s, a flourishing women's movement had emerged, creating collectives, cooperatives, and consciousness raising groups, many of which were organized according to principles of non-hierarchy. Women had developed distinctively "feminist" styles of organization and action, instituting small non-hierarchical groups such as the consciousness raising group, as the cellular structure from which would emerge a national and international movement.

These institutions were designed to give women freedom from particular bodily harms such as rape, battering, and abuse from the male medical establishment. Indeed, projects such as women's health centers, rape crisis centers, and shelters for battered women constituted an institutional expression of the radical feminist demand for freedom from male control of women's bodies.

Yet, in addition to representing a demand for freedom from bodily harm and oppression, there was a tendency within radical feminism to demand the freedom to enjoy the body as a site of liberation, passion, and pleasure. Recognizing the degree to which their sexuality, creativity, and intelligence had been shaped by men, feminists realized that they could rethink their own bodily experience. Women began to create a new aesthetic based on an affirmation of sexuality, intuition, spirituality, art, and health. The arrival of innovative forms of "women's" literature, music, art, theater, dance, and ritual signaled the construction of a "universal woman" who could forge a new identity based on self-love, power, and creativity.

The implicit ecological impulse within radical feminist body politics, then, reflected an emerging social, rather than individualistic, desire for a quality of everyday life infused with bodily freedom, safety, and pleasure for 'all women'. Citing 'patriarchy', or male dominated hierarchy, as the cause of women's oppression, radical feminism sought to establish a new set of cultural practices defined in opposition to what women often described as a body-hating society. Within this implicit 'desire for nature', stood a demand for more than abstract values of 'freedom' and 'justice' that marked many of the student movements of the New Left. Instead, we see an attempt to ground questions of freedom in everyday social relationships and cultural practices that reflected values of collectivity, sensuality, health, and self-determination.

The Disembodied Body: The Emergence Of Cultural Feminism

It is here, however, that the social desire for a new embodied sensibility took a risky turn. Moving from concrete issues of health, safety, and institutional structure to more abstract questions of cultural practice and meaning, radical feminism ventured into the pleasurable yet problematic realm of the symbolic. Questions of how to represent new understandings and practices such as health and spirituality, questions of how to symbolically unify 'women' into a 'universal' category that would 'stand for' the cultural feminist subject, became paramount as a movement of predominantly white, middle-class women looked to 'other' cultures for inspiration. These 'cultural feminists' attempted to represent new embodied cultural practices of their own everyday lives by deploying new symbols, meanings, and images that they often 'borrowed' from the symbols, times, and places of other cultures.[2]

Rejecting patriarchal and hierarchical approaches to spirituality, medicine, and aesthetics, radical cultural feminists sought practices intended to empower 'all women'. This search for new cultural practices was again marked by an ecological sensibility as feminists turned to 'nature based' cultures that had their roots in pagan, Neolithic, Eastern, indigenous, Native American, and African traditions. However, this turn to the 'old' to reconstruct the 'new' is often characterized by the tendency toward abstraction and romanticization: the desire for an idealized 'golden age' expressed by women who drew inspiration from cultures of the past believed to be free of gendered hierarchy and ecological injustice.

The failure of many radical feminists to problematize the process by which they cultivated symbols to represent and routinize feminist nature-based cultural practices contributed to the problem of essentialism within 'cultural feminism'. That many women of color did not identify with symbols that white women deemed 'universal' women's symbols, and that many indigenous women criticized the appropriation by white women of symbols and practices of their own cultures, reflects the failure of white radical feminists to be sufficiently self-conscious about the social and political contingencies that constrain the ways in which feminists reconstruct past and present categories of gender and culture. Indeed, in Audre Lorde's essay, "An Open Letter to Mary Daly," Lorde inquired why Daly used symbols from pre-capitalist Western Europe to represent an empowering cultural image of 'women'. Lorde asked herself, "why doesn't Mary deal with Afrekete as an example? Why are her goddess images only white, western European, judeo-christian? Where was Afrekete, Yemanje, Oyo, and Mawulisa?"[3]

The radical potential of early feminism, then, was undercut by problems of symbolic representation and cultural practice; problems that reflected deeper issues of institutional racism within the movement. By the mid-1980s, radical women of color had confronted the feminist movement on its inadequate analysis of race, class, and ethnicity, illustrating that the 'unified body' of the body politic mirrored only a small minority of the diverse world body of women. The 1987 publication of the anthology *"This Bridge Called My Back,"* edited by Gloria Anzaldua and Cherri Moraga, signaled an era in which women of color transformed the politics of representation forever. *This Bridge* created a forum in which women who previously had no voice in the feminist movement were able to write critically about issues of race, gender, culture, and power.[4]

Other feminist writers of color during this time challenged as well an analytical framework predicated on a binary between domestic and public deployed by white feminists at the time. This understanding of a "domestic/public split" can be traced back to Simone de Beauvoir's 1958 publication of *The Second Sex*, which rooted the universal cause of women's oppression to be their ghettoization within the 'embodied' realm of domestic sphere and their exclusion from the public sphere of work and culture. For de Beauvoir, women's liberation would follow the liberation of women from this embodied domestic realm into the public sphere enjoyed by men.

As bell hooks articulated in her 1984 essay "Rethinking the Nature of Work," the idea that "all women" would be liberated by moving beyond the domestic sphere was based on a classist and racist set of assumptions:

> Attitudes towards work in much feminist writings reflect bourgeois class biases. Middle-class women shaping feminist thought assumed that the most pressing problem for women was the need to get outside the home and work—to cease being "just" housewives...They were so blinded by their own experiences that they ignored the fact that a vast majority of women were already working outside the home, working in jobs that neither liberated them from dependence on men nor made them economically self-sufficient. [5]

In this way, questions of race and class complexified previously universal notions of gender and the body tied to the feminist project. No longer was "woman" a universal subject trapped within a timeless domestic sphere, the escape from which would provide universal liberation. Indeed, for poor women of color who had been "working" outside the home for centuries, there had clearly been no such liberation.

As the writers in *This Bridge* illustrated, the body politic, originally intended to counter the abstract politics of men in the New Left, had given rise to a cultural feminism that presented a new set of abstractions. Just as the New Left had organized its political agenda within liberal and universal categories of 'man', and 'justice' generalized from a particular privileged group of white men, the radical feminist movement had organized its agenda around universal categories of 'woman' and "domesticity" generalized from a privileged group of white women. By failing to sufficiently articulate issues of race, class, and ethnicity, radical feminists were unable to fully clarify the many social factors that determine the particular ways in which women experience and resist oppression. Audre Lorde, again, in her letter to Mary Daly, questioned Daly on the white bias surrounding her body politics, stating:

> You fail to recognize that, as women, there are (vital) differences which we do not all share. For instance, breast cancer; three times the number of unnecessary eventrations, hysterectomies and sterilizations as for white women; three times as many chances of being raped,

murdered, or assaulted as exist for white women. These are statistical facts, not coincidences nor paranoid fantasies.[6]

Audre Lorde was one of the first radical feminists to bring to body politics an understanding of the relationship between race, health, class, and gender. In her ground breaking work, *The Cancer Journals*, Lorde examined the specific social context in which she had been exposed to toxins at home and at work.[7] In addition, she articulated the specific social contexts in which she faced her own medical crises and recovery. Lorde's perspective anticipated the struggles of women of color in the environmental justice movement of the 1980s; a struggle to bring questions of race and class into an ecologically oriented body politic.

Thus the 'body politics', which offered a potential 'organic' ground for radical feminism, was constrained by a tendency toward abstraction and romanticization. Indeed, degrees of immediacy and historicity were lost in the translation as white women began to extrapolate from their own lives a politics of representation that often either appropriated or excluded the experience of women of color. And as we shall see, this problem of how to engender new meanings surrounding categories of non-hierarchy, body, gender, and nature, persisted as a nascent desire for nature continued to emerge within radical feminism.

Yet despite these limitations, by framing issues of health, sexual freedom, rape, and battering, as political issues, radical feminists began to move toward a social, rather than individualistic, desire for nature, expressing a collective desire for a more healthful, pleasurable, and "natural" expression of everyday life free from social oppression. In turn, the nascent anarchist impulse that marked the cooperative structure of feminist organizations speaks to the revolutionary potential within feminist body politics.

Body-Ecology: The Emergence Of Ecofeminism

To explore the movement of radical feminist body politics into an explicit desire for nature, we will return briefly to the earlier days of the movement. Here, once again, we witness a set of mostly white, middle-class activists for whom ecological questions will represent an attempt to make sense out of abstract understandings of categories of nature and gender: understandings that will reflect their own identities.

The WITCH movement represents one of the first feminist actions that expressed an explicit ecological sensibility. At this time, feminists began to articulate moments of resonance between the idea of a new 'embodied' political culture and the culture of witches in pagan Europe hundreds of years ago. Beginning on Halloween, 1968, radical feminists formed a series of autonomous 'covens' across the country. The group was explicitly non-hierarchical, and their style was theatrical, humorous, and passionately strident. They expressed a brilliance of wit in their ever-changing acronyms ranging from Women's International Terrorist Conspiracy from Hell, and Women Infuriated at Taking Care of Hoodlums, to Women Interested in Toppling Consumption Holidays. A coven in New York City leafleted a statement that would anticipate later ecofeminist writings:

> WITCH is an all-woman Everything. It's theater, revolution, magic, terror, joy, garlic flowers, spells. It's an awareness that witches and Gypsies were the original guerrillas and resistance fighters against oppression—particularly the oppression of women—down through the ages. Witches have always been women who dared to be: groovy, courageous, aggressive, intelligent, non-conformist, explorative, curious, independent, sexually liberated, revolutionary. (This possibly explains why nine million of them have been burned.) Witches were the first Friendly Heads and Dealers, the first birth-control practitioners and abortionists, the first alchemists (turn dross into gold and you devalue the whole idea of money!). They bowed to no man, being the living remnants of the oldest culture of all—one in which men and women were equal sharers in a truly cooperative society, before the death-dealing sexual, economic and spiritual repression of the Imperialist Phallic Society took over and began to destroy nature and human society. [8]

In one action, a coven in Washington D.C. 'hexed' the United Fruit Company because of their "oppressive policy on the Third World and on secretaries in its offices at home." A leaflet distributed at the demonstration contained the spell:

> Bananas and rifles, sugar and death,
> war for profit, tarantulas' breath,
> United Fruit makes lots of loot,
> the CIA is in its boot.[9]

As early as 1969, women were beginning to bring together an analysis of militarism, capitalism, sexism, and colonialism that was regarded as destroying "nature and human society." In this action we see a light-hearted, yet significant, 'backward-looking' impulse that will mark both cultural feminism and later forms of 'cultural' ecofeminism. The witty and romantic appeal to a 'witch culture' of the past represents an attempt by a group of mainly white suburban youth to invoke the idea of an era that was more cooperative and ecological.

In 1978, Susan Griffin wrote *Woman and Nature*,[10] a book-length prose poem that juxtaposed objectified representations of women with managerial writings about plant and animal 'nature'. Griffin's book, which soon became part of an emerging radical feminist/ecological cannon, was influential in revealing the socially constructed correspondence between ideas of 'woman' and 'nature' within capitalist patriarchy. In 1980, Carolyn Merchant published an important feminist perspective on the scientific revolution, further contributing to this newly developing feminist ecological literature. Merchant's book, *The Death of Nature*, discussed the historical relationship between capitalism, modern science, and women's oppression.[11] Merchant, a socialist feminist, articulated how patriarchy and capitalism functioned together to control both 'woman' and 'nature'.

During these years, the body politic expanded to address not only understandings of women's physical survival and vitality, but ideas of 'global' survival in general. Once early feminists asserted that 'patriarchy' had invaded their very bodies, it wasn't a big leap for them to assert that the same system had invaded the rest of the natural world as well. However, the ways in which women articulated the causes of ecological problems varied immensely. In both the WITCH movement and in the writings of Merchant, there is a critique of capitalism that names capitalism in particular, not just 'patriarchy' in general, as a primary cause of ecological malaise. In contrast, Susan Griffin's book displays the 'universalizing tendency' that marked much of 1970s radical feminism; a tendency to identify 'man' in the abstract as the cause of ecological injustice:

> The fact that man does not consider himself a part of nature, but indeed considers himself superior to matter, seemed to me to gain significance when placed against man's attitude that woman is both inferior to him and closer to nature. Hence this book called *Woman and Nature* grew.[12]

Yet while Griffin reproduces the essentialist tendency that had emerged within cultural feminism, she does extend a radical feminist analysis of social hierarchy to an exploration of ecological concerns. According to Griffin, problems of sexism and ecological malaise are caused by men who regard themselves as 'superior to', rather than 'part of', nature. Thus in *Woman and Nature*, Griffin suggests the idea of a potentially complementary relationship between society and nature, given the right social conditions.

By the early eighties, feminists began to define the organic sensibility latent within radical feminist body politics in more explicitly ecological terms.[13] Radical feminists began to develop the idea of a time that was prior to social and ecological injustice, a time in which 'women' had more power and control over their everyday relationships with each other and with nature. Women began to cultivate a desire for nature that conveyed a yearning for a more cooperative way of life free of sexism and ecological degradation.

THE ANTI-NUClEAR MOVEMENT ANd EcofEMiNisT ACTivisM Of THE EARly 1980s: BRiNqiNq TOqETHER PEACE ANd Ecoloqy

During this time, another movement had been gaining steam. In the seventies, anti-nuclear activism emerged as one of the most potent political forces within the New Left. In particular, the nuclear issue brought together both radical feminists involved in feminist peace politics and women interested in ecology. While nuclear *militarism* resonated with concerns of feminists peace activists, nuclear *power* became the focus for feminists concerned with problems of ecology and health. Continuing to utilize the domestic/public framework introduced in the 1960s, many radical feminists extended their critique of "domestic" acts of male violence such as rape and battering, to include a critique of "public" and institutional acts of male violence such as militarism. It was in this context that many women began to make connections between the domination of women in the domestic sphere (within personal, sexual relationships) and the destruction of the natural world by public institutions such as the military and the nuclear industry.

The feminist peace movement, emerging out of radical feminism and the civil rights and anti-war movements, greatly informed a newly emerging ecofeminist activism. Inspired by the philosophy of anti-racist peace activists such as Barbara Deming, feminists had been developing an anti-militarist movement in response to mounting U.S. aggression. Learning of the nuclear testing in Nevada in the fifties and the subsequent rise in birth defects and gynecological cancers, they also discovered the current problem of nuclear waste for which there was no safe means of disposal. And while appreciating the ecological implications of nuclear energy, feminists also addressed the military implications of an industry that produced plutonium necessary for nuclear warheads. The issues of militarism, male violence, and ecology came together to form a truly ecological, broad-based body politic.

In 1980, the crisis at the nuclear reactor on Three Mile Island served as the catalyst for a beginning of ecofeminist direct action. This first major ecofeminist event was initiated by feminist activists Ynestra King and Celeste Wesson during an interview on WBAI radio in New York in which they discussed the crisis from a specifically ecofeminist perspective. The following April, King and Wesson, along with a group of other feminist, peace, and environmental activists, organized "The Conference on Women and Life on Earth: Ecofeminism in the 80s" in which 800 women gathered in Amherst, Massachusetts to address the nuclear question. Many of the conference organizers and attendees identified as social anarchists who had been involved in the anti-nuclear movement.

Out of this conference emerged an ecofeminist network that, in 1981, planned the first ecofeminist action: the "Women's Pentagon Action" (WPA) in which 3,000 women participated in a massive theatrical ecofeminist demonstration in Washington D.C. The WPA was an ecofeminist and anti-militarist action whose "Unity Statement," written collectively and arranged by Grace Paley, tied together issues of feminism, capitalism, ecology, anti-racism, and anti-militarism:

> With that sense, that ecological right, we oppose the financial connections between the Pentagon and the multinational corporations and banks that the Pentagon serves. Those connections are made of gold and oil. We are made of blood and bone, we are made of the sweet and finite resource, water. We will not allow these violent games to continue. If we are here in our stubborn thousands today, we will certainly return in the hundreds of thousands in the months and years to come.[14]

In the first WPA action (there was another the following year), activists used a style reminiscent of the WITCH actions, circling the Pentagon to express rage, sadness, and fear about the history of male violence by performing street theater on the Pentagon's steps. While the WPAs echoed the sensibility of the WITCH movement, they also echoed the domestic sensibility of an earlier anti-nuclear movement of 1962, known as the "Women's Strike for Peace" movement, in which women from across the country, identifying as 'mothers' (rather than as feminists) demonstrated against the nuclear testing that had taken place in the fifties.

Whereas radical feminism had been often criticized for espousing an anti-mother sentiment (traced back to de Beauvoir's assertion of women's need to transcend the maternal activities associated with the domestic sphere), early ecofeminists reversed de Beauvoir's assertion, arguing instead that women must restore value to the roles of mothering and nurturing. This motherist sensibility (often blamed for creating yet another romantic essentialism) was translated into the creation of a form of direct action that came to be associated with ecofeminist actions in the future. Blending both 'witchy' and 'motherist' sensibilities, the WPAs created a new kind of distinctively ecofeminist aesthetics. At the WPAs, women wove webs of yarn containing symbols of mothers' everyday lives, such as aprons, clothespins, photographs of children as well as artifacts from women's everyday lives around fences, doors and missile sites as described by Ynestra King:

> We create an iconography designed to bring people to life—parading with enormous puppets, quilting scenes from everyday life, weaving the doors of the Pentagon closed with brilliantly colored yarn, waltzing around police barricades, shaking down fences, spray-painting runways, placing photos of beloved places in nature

and children woven in the miles of fencing around military installations, wearing flowers and brilliant colors as we face into the gray and khaki of militarism, opposing machines with hand-crafted alternatives. [15]

By reversing (yet reproducing) the domestic/public split as an analytical framework, the WPA began to counter the values of capitalist consumerism and state militarism by expressing a new revalorization of the everyday life of the domestic sphere.

By 1981, an international ecofeminist network had emerged. Ecofeminism, with its analysis of the interconnectedness of oppressions and its insistence on the need for international dialogue, provided a global forum for addressing women's social and ecological crises. In response to this 'missile crisis', a group of British peace and ecology activists, along with the recently established group, 'Women and Life on Earth' in England, created the Greenham Common Peace Encampment at the military base located there. At the time, Greenham represented an ongoing international direct action, a demonstration of women's work of everyday survival in a patriarchal nuclear age. Setting up camp outside the gates of the base, women lived in tents and shelters and were re-evicted each morning by the military police. Subsequently, in solidarity with Greenham, women in the U.S. founded the Seneca Women's Peace Encampment in Seneca Falls, NY, to protest cruise missiles that were positioned to leave Seneca for Europe.[16]

Finally, in the mid-1980s, a group of ecofeminists began to specifically address issues of race and class in relationship to the ecofeminist project. Initiated in 1984, the WomanEarth Feminist Peace Institute was founded by a group of radical women of color, ecofeminists, and feminist peace activists including Ynestra King, Gwyn Kirk, Barbara Smith, Rachel Bagby, Luisah Teish, and Starhawk, who came together to create a multi-racial, multi-cultural forum in which women could discuss issues of race, gender, class, peace, spirituality, and ecology. Following the suggestion of Barbara Smith, WomanEarth became the first feminist institute to be organized around the principle of racial parity, giving equal voice, participation, and leadership to both women of color and white women.[17]

While WomanEarth sought to become an educational and political institute that could provide a base for an ecofeminist movement, internal struggles within the organizing group regarding race and class privilege, in addition to financial pressures, led to the eventual dissolution of the project in 1989. As Noël Sturgeon points out, however, WomanEarth still serves as an example of a moment in ecofeminist history in which white ecofeminists placed questions of racial privilege and power at the center of their political agenda. The commitment that ecofeminists brought to this project was reflected in WomanEarth's conference "Reconstituting Feminist Peace Politics" held in Amherst, MA, in June of 1986, a conference in which fifty women (half women of color, half women of European descent) met to discuss a range of issues relating to questions of race, class, and feminist peace politics. WomanEarth signaled an important shift within ecofeminism. Responding to critiques of racism within the feminist movement as a whole in the mid-1980s, women such as King understood that ecofeminism had to prioritize the question of racism if the movement was to achieve political validity and integrity.[18]

WomanEarth, as an ecofeminist project, emerged out of radical feminist body politics that sought to particularize the general question of ecology by addressing issues of 'nature' along with those of gender and social justice. Initially, the nuclear issue brought out the most concrete, social, and historical dimensions of the "nature question" within ecofeminism. Departing from mainstream environmentalism's tendency to privilege abstract notions of a pristine and "people-less" wilderness to be protected, these early ecofeminist activists generally expressed their "desire for nature" by showing the concrete connections between public and domestic acts of militarism and male violence, pointing to the ecological and social implications of such issues. Again, although early ecofeminist activism tended to reproduce the problematic domestic/public framework, they were able to ground their politics in a social and material analysis of ecological questions.

Thus, in the early 1980s, radical feminism had given rise to an increasingly social approach to ecological questions that grew out of a body politics grounded in the concrete dimensions of women's everyday lives. This body politics was predicated upon the ability of radical feminists to link questions such as health and sexuality to systems of male dominated hierarchy, reflecting a nascent, and sometimes explicit, anarchist impulse. And as we have seen, this nascent anarchism within body politics finds expression within early ecofeminist claims regarding the connection between ecological degradation and questions of social domination and oppression in general.

Social Ecology And Ecofeminism

At this point in the narrative, it would be helpful to take a few steps back to explore a key political and theoretical context in which Ynestra King, a major figure in the early years of ecofeminist activity, developed ecofeminist theory and activism. King's approach to ecological theory and politics both informed, and was formed by, another desire for nature that unfolded simultaneously with the radical feminist movement. That desire for nature is social ecology.

Social ecology is a branch of the radical ecology movement that surfaced in the U.S. during the 1960s. Since its inception, social ecology has played a major role in shaping radical ecological politics both in the U.S. and abroad by pushing ecological discussion in a social anarchist direction to include critiques of capitalism, the state, and all forms of social and political hierarchy. Beginning in the early 1960s, Murray Bookchin, the theorist primarily associated with the theory, began to examine the social and political origins of ecological problems from a leftist perspective. While offering a philosophical and historical analysis of the relationship between society and nature, social ecology is praxis-based, calling not only for direct action, but for a reconstructive vision of a confederation of communities engaged in direct democracy and municipalized economics.

While an ecological sensibility emerged within the body politics of radical feminism in the 1960s and '70s, a nascent feminist sensibility surfaced within social ecology. The common denominator that led both radical feminists and social ecology to make the connection between ecology and feminism can be traced back to the anarchist impulse within both theories. While early feminist analysis of hierarchy led to a critique of the 'patriarchal' project to dominate nature, the social and ecological analysis of hierarchy led to a critique of systems of male domination.

Inspired by the newly emerging radical feminist movement, Bookchin too, saw in feminism, as he saw in ecology, the potential for a movement that was general enough to include, yet not be limited to, economic concerns. Like others, Bookchin saw feminism as potentially one of the "great issues" that, like ecology, democracy, and urbanization, could bring to the revolutionary struggle those who faced hierarchical as well as class oppression.[19] He recognized in feminism the potential for a trans-class movement that could lead to an anti-hierarchical position that could ultimately challenge capitalism.

In 1978, the Institute for Social Ecology (ISE), which Bookchin co-founded in 1974, invited Ynestra King to develop what would become the first curriculum in a feminist approach to ecology, thus coining the term ecofeminism.[20] As there were not yet any explicitly ecofeminist writings, King created the first ecofeminist curriculum which reviewed essays written by theorists including liberal, socialist, radical feminist, and anti-militarist thinkers, as well as feminist anthropologists and feminist philosophers of science. Through a critical reading of these essays, King explored the evolution of feminist thinking from the first to the second wave, looking at moments of liberalism, rationalism, and essentialism within the different strands of feminist theory, examining their implications for ecological theory and feminist peace politics.

Bringing together insights gleaned from both social ecology and feminist epistemology, King developed a way to rethink the self/other relationship central to both ecology and feminism. In particular, King drew from feminist theorists such as Nancy Chodorow, Gayle Rubin, and Sherry Ortner, examining the historical implications of the Western nature/culture dichotomy for the construction of gender.

For King, the woman/nature analogy was a social, rather than biological, construction that she sought to historicize and appropriate as a way to develop a feminist critique of the epistemological foundations of Western society. According to King, this analogy was directly linked to a "nature/culture split" which was in turn, tied to the domestic/public dichotomy discussed by white feminists during the late 1960s and early '70s.[21] Again, departing from de Beauvoir, King called for women to analyze the historical construction of that dichotomy as a way to understand men's alienation from "domestic" realms of nature and the body, rather than for women to join men in the project of "transcendence" over nature. However the failure of King (and of many white feminists at the time) to problematize the domestic/public split itself, left early ecofeminist theory vulnerable to critiques of essentialism that continue today. As already stated, the tendency among white feminists during those years to focus on the domestic/public dichotomy reflected unexamined assumptions regarding the universality of the structural causes of women's subordination. Again, as theorists such as bell hooks pointed out, poor women of color in the U.S. had always been forced into the "public" sphere of work—without "transcending" their oppression as women.

Yet while retaining this problematic domestic/public framework, King's approach to ecofeminism was profoundly radical in a variety of ways. Social ecology had provided an explicitly revolutionary, anarchist, and ecological lens through which King analyzed questions regarding objectivity raised by feminist psychoanalytic theorists, scientists, and anthropologists. Offering a way to 'ecologize' the Hegelian dialectic between self and other, social ecology articulated the need for society to create a relationship with the rest of the natural world marked by degrees of cooperation, complementary, and ever greater levels of freedom. Social ecology's discussion of 'unity in diversity' also provided a way to reconcile the relationship between self and other by articulating the possibility for recognizing both the differences and connections between organic phenomena. Within the 'ecologized' dialectic of social ecology, the self could be both related to, and distinct from, the other.

King drew out the feminist implications of social ecology, exploring non-hierarchical and anarchic ways of approaching self/other relationships in domains of political and ecological organizing and theory. In addition to teaching at the ISE, King went on to create the first body of writing to be called explicitly "ecofeminist," creating an innovative synthesis of theories including social ecology, radical feminist body politics, feminist critiques of science, feminist peace politics, and critical theory.[22] Yet while King sought to integrate feminist and social ecological theory, she articulated in turn, the need for a feminist dimension to the theory of social ecology:

> The perspective that self-consciously attempts to integrate both biological and social aspects of the relationship between human beings and their environment is known as *social ecology*...while this analysis is useful, social ecology without feminism is incomplete. Feminism grounds this critique of domination by identifying the prototype of other forms of domination: that of man over woman.[23]

In this way, King drew out the feminist implications of social ecology, exploring new ways of examining the relationship between systems of male domination and ecological crises in general from a perspective informed by social anarchism. Although feminists such as those in the WITCH collective were drawing similar connections between oppressions almost a decade earlier, King made the articulations between forms of social hierarchy explicit, demonstrating their relationship to ecological injustice.

King's grounding in anarchist theory and social ecology allowed her to avoid many of the epistemological traps into which feminists fell during those years. Through a social ecological critique of hierarchy, she recognized the need to abolish all forms of oppression, while emphasizing as well, the potential for political collaboration between women of different class, race, and ethnic backgrounds. King's key role in establishing WomanEarth, as well her participation in international feminist forums such as the United Nations Conference on Women in Nairobi in 1985, reflect her epistemological sensitivity to questions of difference as well as her anarchistic appreciation of the need to simultaneously fight against all forms of hierarchy and oppression.

King's ecofeminism did more than just recognize the importance of making connections between different forms of social and ecological injustice: It recognized the importance of making connections between different *women* all over the world to counter these interconnected crises. Repeatedly in her writings, King expressed the need to create face-to-face dialogue between women, both internationally and cross-culturally within the United States, to create unified anti-racist strategies to address women's diverse struggles for social and ecological justice.

Ecofeminism, Environmental Justice, And International Environmentalism

To fully appreciate the historical distinctiveness of King's participation in multicultural and anti-racist projects such as WomanEarth, we must locate it within a larger history of both the feminist and ecology movements of the mid-1980s. As a mostly white feminist movement was being challenged regarding problems of racism and essentialism, the ecology movement was confronted on its exclusion of the concerns and participation of communities of color. WomanEarth most particularly reflected the simultaneity of these challenges as white women active in both feminist and ecology movements began to prioritize the issue of race within both the feminist and ecological agenda.

While WomanEarth was forming, two other forums emerged in which women addressed questions of race, culture, class, and ecology: the environmental justice movement and the movement surrounding feminist international environmental politics. I include a discussion of these movements as a way to depict the wider, politicized climate of the environmental movement in which ecofeminism was located in the mid-1980s to better contextualize concerns faced by ecofeminists during this time.

During the mid 1980s, the grassroots anti-toxics movement, which had previously been composed of mostly white communities fighting toxic dumping, also began to undergo a transformation. Activists of color who had fought for decades against environmental injustices that targeted their communities throughout the U.S., began to take leadership in this movement, and within the wider environmental movement, linking questions of social, political, and economic justice to the ecological question. They began to recast issues previously regarded as 'community' or 'social' problems in 'ecological' terms. In so doing, they appropriated an ecological discourse from which they had been marginalized.

The anti-racist wing of the environmental justice movement emerged in response to the marginalization of people of color from the mainly white ecological millieu. To mainstream white environmentalists, community-based struggles of activists of color are often understood as 'social' rather than 'environmental'.[24] Ongoing attempts within poor communities of color to secure services such as paved streets, sewers, indoor plumbing in addition to struggles for a pleasurable quality of everyday life, have been largely ignored by mainstream environmentalists as such issues often fall outside of, or between, the boundary that separates 'the city' and 'the country'; a boundary that exists within the Euro-American environmental imagination. In this way, then, neither the cityscape nor the poor rural community in which activists of color work to achieve quality of life, fit white categories of 'social' and 'environmental'. Indeed, according to activist and theorist Dorceta E. Taylor, the myth that people of color are unconcerned with environmental issues is allowed to continue due to the way that white mainstream environmentalists frame and strategically address ecological problems.[25]

However, by the late 1980's an environmental coalition of activists emerged from within the African American, Native American, Puerto Rican, Latino, and Asian and Pacific Islander communities: a coalition to fight environmental racism. Environmental racism includes the official sanctioning of polluting industries, poisons, and pollutants in communities of color in addition to the exclusion of people of color from environmental policy making, regulatory bodies, and from mainstream environmental groups. Unlike mainstream environmentalism or deep ecology, the struggle against environmental racism does not historically emerge from an abstract or romantic desire for nature expressed as a yearning to 'protect' a pre-social idea of nature, but from an historical appreciation of the inseparable conditions of ecological and social injustice.

Unlike early ecofeminist theory that emerged out of the analytical framework of domestic/public or nature/culture, the environmental justice movement tended to deploy categories defined in terms of race, class, and culture. For activists in the environmental justice movement, environmental problems are not seen to be the result of man's alienation from an embodied, domestic sphere identified with women. Instead, environmental injustice is seen to be the consequence of a specifically Western, racist, and capitalist society that has constructed itself at the ecological and cultural expense of poor communities of color.

Thus, in the movement for environmental justice, we see another expression of the desire for nature, a desire for ecological integrity that reflects yet another set of identities and situations. Often identifying as members of indigenous cultures or communities of color struggling for survival, rather than as "feminists" (a term emerging out of white middle-class context), a new wave of women leaders arose during the 1980s, changing the ecological landscape in the U.S. Over the past ten years, women such as Winona La Duke, Peggy Dye, Dorceta E. Taylor, Vernice Miller, and Cynthia Hamilton have emerged as internationally recognized leaders in the struggle to end environmental injustice. According to Cynthia Hamilton:

> Women often play a primary part in community action because it is about things they know best. Minority women in several urban areas have found themselves part of a new radical core as the new wave of environmental action, precipitated by the irrationalities of capital intensive growth, has catapulted them forward. These individuals are responding not to nature in the abstract but to the threat to their homes and to the health of their children.[26]

Women active in struggles against environmental racism have particularized the ecological question with a politics grounded in an analysis of history, capitalism, and racism. During a time when many deep ecologists and mainstream environmentalists rarely speak of capitalism as a factor in ecological and social devastation (referring instead to euphemisms such as 'technology' 'modern society', or 'industrial society'), environmental justice activists, such as Cynthia Hamilton, have consistently named capitalism as a primary force behind ecological and social injustice.

Women in the environmental justice movement became a source of inspiration to white ecofeminists who, by the mid-1980s, were at a loss for how to reconstitute an activist base for the movement. Indeed, in contrast to the ecofeminist movement which was constituted in national anti-militarist campaigns, women involved in the fight for environmental justice were engaged in community based, struggles for cultural and ecological justice tied to everyday issues ranging from land rights to toxic waste. Yet while white ecologists have often been drawn to the work of environmental justice activists such as Winona La Duke, often seeking their endorsement of the movement, ecofeminism *per se* has not held significant appeal or relevance to women engaged in local struggles for community and cultural survival. Women in these movements tend to identify as 'community' or 'environmental' rather than 'feminist' activists. Though the two groups are primarily led by women engaged in ecological concerns, there has been little overlap between environmental justice organizing and ecofeminism.

In turn, the continuing segregation of communities of color and white communities, combined with unresolved tendencies toward white bias within feminist theory, have greatly impeded the formation of coalitions between white ecofeminists and women of color active in the environmental justice movement. Within this context, WomanEarth represented an important moment in ecofeminist history. Recognizing that a multi-cultural, multi-racial project such as WomanEarth would require intentional and careful planning involving both white women and women of color from the beginning stages, WomanEarth signaled an attempt by ecofeminists to address racial constraints that hindered the movement from fulfilling its potential. Rare moments such as WomanEarth reflect the racialized context of ecological politics in the U.S., complexifying abstract notions of 'woman' and 'nature' that lingered within ecofeminist theory during these years.

There has been considerably greater overlap between ecofeminists in the North and women in South engaged in development discourse. This coming together was originally facilitated by two international conferences sponsored by the United Nations (UN) Decade for Women designed to provide forums in which women could meet to discuss their economic and social status in an international setting. Launched in 1975, the Decade for Women intended to trace the improving status of poor women in the Third World during the ten years of a UN funded development campaign. However, the research instead revealed that the lives of many poor women had actually worsened during the ten years, as women had to bear not only the declining economic conditions brought on by a new phase of neo-colonialism, but the ongoing burdens of sexism as well.[27]

At the end of the Decade, in 1985, the UN sponsored the Second UN Conference on Women in Nairobi, stimulating unprecedented discussion between northern and southern feminist activists, shedding light on the global, diverse, and complex nature of women's approaches to social and ecological questions. The Nairobi conference signaled the beginning of a new international phase of feminist activism and dialogue that, like the publication of *This Bridge*, began to challenge universal categories of gender, as well as domestic/public binaries, that marked white ecofeminism in the U.S. In addition, as women in the South spoke publicly about multiple issues of globalization, cultural identity, and development, they began to challenge essentialist understandings of the monolithic "Third World Woman" or "indigenous woman" that were embedded within white feminism of the 1980s.

For many poor women in Third World situations, discussions of "development" reflect a desire for ecological integrity, that in turn, are born out of a particular set of identities and situations. For many in the South, the desire for "nature" is rooted in an analysis and critique of colonialism, global capital, sexism, and environmental policy—rather than out of a nature/culture dualism. Within such discussions, "nature" itself is a contentious ground owned and controlled by international regulatory agencies, development agencies, and trade agreements. In turn, "nature" also often represents a set of agricultural, economic, medicinal, spiritual, and cultural practices based on local knowledge built up over generations.

For women in subsistence economies, ecology often represents the day-to-day articulations between an encroaching global capitalist economy, governmental formations, and traditional organic cultural symbolic practices. In turn, for many poor southern women undergoing processes of proletarianization within newly emerging industrialized contexts, ecological issues mean not only poisoned water and air, but toxic work places where women are exposed to harmful chemicals, over-work, and under-pay which keep women in a continual state of stress and poverty.

Through international dialogue, women addressing issues surrounding development began to articulate a "global feminism" that brings together the economic, cultural, and ecological insights of women in both the North and South. Vandana Shiva, one of the few environmental activists from the South to identify with the term 'ecofeminism', has emerged as a major voice in global feminist forums. In her work over the last fifteen years, Shiva has articulated the struggles of women in rural India to resist colonial policies of deforestation, agriculture, and land use. In particular, as a socialist ecofeminist, Shiva has been instrumental in elucidating issues relating to biotechnology and seed patenting, tying issues of biotechnology to the larger struggle between neo-colonialism, global capital, ecological sustainability, and women's local knowledge.[28]

The emergence of post-colonial feminist discussion in the mid-1980s brought U.S. ecofeminists engaged in such forums into a transnational feminist movement. Ecofeminists have assumed leadership in international forums such as the Women's Environment and Development Organization (WEDO) which sponsored the World Women's Congress for a Healthy Planet in November of 1991. While WEDO is not an explicitly ecofeminist organization, a distinct ecofeminist perspective is visible within their literature that still emphasizes the woman/nature dichotomy and the question of peace. Indeed, WEDO's Declaration of Interdependence of 1989 is reminiscent of the Women's Pentagon Action's Unity Statement almost a decade before:

> It is our belief that man's dominion over nature parallels the subjugation of women in many societies, denying them sovereignty over their lives and bodies. Until all societies truly value women and the environment, their joint degradation will continue...Women's views on economic justice, human rights, reproduction and the achievement of peace must be heard at local, national, and international forums, wherever policies are made that could affect the future of life on earth. Partnership among all peoples is essential for the survival of the planet.

Yet while retaining some of the analytical categories of its earlier "anti-militarist" days, U.S. ecofeminists in international forums such as WEDO have sought to link questions of nature to issues of gender, social justice, and health, thus expressing a desire for nature that tends to be socially, rather than individually, based. Again, when we compare WEDO's Declaration to anti-humanist statements written by many in the deep ecology movement during the late 1980s, we can better appreciate the significance of ecofeminist attempts to raise questions of "economic justice, human rights, reproduction, and the achievement of peace" in relation to the question of ecology.

The shift from an ecofeminism derived from a U.S. based anti-militarist movement to a transnational ecofeminism focused on questions of development, complexified ecofeminist theory, both broadening and grounding the idea of the ecological subject. As poor women in the South inscribed issues of development, colonialism, and globalization as 'ecological', they unsettled universal assumptions often built into northern ecofeminists' "desires for nature."

U.S. Ecofeminism Of The Late 1980s And Beyond

While ecofeminists from the U.S. participated in international feminist forums during the mid-1980s, an autonomous ecofeminist movement in their own country began to wind down. The early years of U.S. ecofeminist activity were for many the 'high point' of the movement's history. Punctuated by the Women and Life on Earth Conference, WPAs, Seneca Peace Encampment, WomanEarth, and an array of local actions in the Northeast and throughout the country, these short years in the early 1980s were a time in which U.S. ecofeminism was particularly rooted in an activist tradition originally constituted by the New Left.

Indeed, by the late 1980s, although many individual ecofeminists were active in Green movements, struggles for animal rights, and forest defense work, there was little to suggest that autonomous ecofeminist activism would be revived. If ecofeminism did not take to the streets, it took to the many literary and educational forums that would proliferate over the next decade. The bursts of early ecofeminist activity had captured the imaginations of a wide range of activists, students, and scholars interested in feminist critiques of science, environmentalism, animal rights, feminist theology, and feminist philosophy, both within and outside of the academy. By the early 1990s, there were three ecofeminist anthologies, an array of ecofeminist journals, related books, major conferences, workshops and university curricula that helped to further stimulate excitement about ecofeminism.

During this time, some left-oriented feminists noticed a problematic tendency within the movement: its vague relationship to anarchist or leftist politics. The ecofeminism introduced by King at the ISE was linked to a vision of a non-hierarchical, ecological society free of statist and capitalist social relations.[29] The Women's Unity Statement of the WPAs reflected this sentiment by challenging the power of the state and capital through its defamation of the Pentagon, the U.S. government, and multinational corporations.

From a social ecofeminist perspective, an ecofeminist perspective informed by social ecology and social anarchism, the writings that filled the pages of the first two major anthologies on ecofeminism were disappointing indeed. Of the twenty-six chapters of the anthology *Healing the Wounds,* published in 1989, there were only two authors, Vandana Shiva and Ynestra King, who mentioned the words capitalism or the state. Instead, writers pointed to the causes of ecological destruction by appealing to terms such as "technology", "patriarchal rationality", "economic motivation" and "industrialization." For instance, in her introduction to the anthology, Judith Plant describes the causes of ecological destruction to be the result of a man's world:

> [T]he world is rapidly being penetrated, consumed, and destroyed by
> this man's world—spreading across the face of the earth, teasing and
> tempting the last remnants of loving peoples with its modern glass
> beads—televisions and tanks; filling the ears of poor peoples with
> doublespeak about security, only to establish dangerous technology
> on their homelands; voraciously trying to control all that is natural,
> regarding nature as a natural resource to be exploited for the gain of
> a few.[30]

In this passage, Plant points to the effects of, and social relations within, a
market economy by discussing the exploitative "gain of the few." Yet Plant fails
to mediate her discussion of the causes of ecological problems with categories
of race, class, or with an understanding of institutional forms of capitalist and
state power. Instead, she invokes universal notions such as this man's world
(retained from radical feminist theory) that did not help to clarify her political
position.

During this time, some social ecofeminists, along with other ecofeminists,
also began to notice a minor, but notable, romantic tendency within several
ecofeminist writings that made the theory a target for unending, and often
unfair, criticisms of essentialism.[31] The second major ecofeminist anthology,
Reweaving the World (containing essays written in the late 1980s),[32] was
punctuated with several unproblematized essentialisms regarding nature and
culture. For example, in her essay "Ecofeminism: Our Roots and Flowering,"
Charlene Spretnak described "the elemental power of the female"[33] appealing
to an essentialist notion of 'gender'. In turn, while reflecting upon the day on
which she introduced her newborn daughter to the world of nature by
bringing her into the backyard of a Los Angeles hospital, Spretnak conflates
this act with that of ritual practiced by Omaha Indians:

> I introduced her to the pine trees and the plants and the flowers, and
> they to her, and finally to the pearly moon wrapped in a soft haze
> and to the stars. I, knowing nothing then of nature-based religious
> ritual or ecofeminist theory, had felt an impulse for my wondrous
> little child to meet the rest of cosmic society...that experience was so
> disconnected from life in a modern, technocratic society...(that) last
> year when I heard about a ritual of Omaha Indians in which the
> infant is presented to the cosmos, I waxed enthusiastic...but forgot
> completely that *I, too, had once been there*, so effective is our cultural
> denial of nature...[34] (emphasis added)

Spretnak's text demonstrated the problem that surfaced as some ecofeminists
asserted universal notions of 'nature', ritual, and cultural practice. As a
middle-class white woman of Christian heritage, Spretnak described giving

birth to a child in a hospital in an industrialized capitalist society in the U.S. The trees and plants on the hospital grounds to which she introduced her child, represented a 'nature' that had been carefully crafted to convey culturally specific understandings of what kinds of plants, grass, flowers, and 'view' should represent 'nature' within the setting of post-industrial Los Angeles. Yet, despite the multiple layerings of time, place, and culture that produced the hospital and its grounds, Spretnak described her surroundings as part of a universal and essential "there" of the Omaha Indians, to which "she, too," once belonged.

I mention this example not to single out Spretnak, nor to construct an essentialist 'straw ecofeminist', but to point to a tendency that emerged as ecofeminist theory was integrated with particular strands of feminist spirituality during the late 1980s. Trying to 'reach' for the ecological in a well-meaning and spiritual way, several theorists failed to sufficiently problematize categories of 'woman', 'nature', and 'culture'. And, while the early 1990s brought eloquent anti-essentialist critiques by theorists such as Val Plumwood and Karen Warren, a popularized version of ecofeminist spirituality endured. Both within the anti-feminist imaginary of those that wage what Greta Gaard refers to as ecofeminist backlash, and within real instances of essentialist ecofeminism outside of the academy, essentialist ecofeminism still flourishes today.[35]

Although the 1990s have not brought a revival of an autonomous ecofeminist movement in the U.S., the decade has given rise to a promising new wave of ecofeminist activism and scholarship. Ecofeminist critiques of deep ecology, initiated in the late 1980s, raised awareness of sexism within such organizations as Earth First! and within forest defense work, signaling increased participation by ecofeminists within such movements. In turn, ecofeminists such as Greta Gaard and Marti Kheel, engaged in animal rights activism, broadened the discussion to include crucial insights into the social and cultural contexts surrounding issues such as vegetarianism and hunting.[36] Within feminist philosophy, ecofeminists such as Val Plumwood and Karen Warren made significant strides in addressing and transcending problems of essentialism within the theory. And quite recently, there have emerged thoughtful and critical discussions of ecofeminist history by ecofeminists such as Greta Gaard, Noël Sturgeon, and Chris Cuomo, ushering in a new era of self-reflexivity by activists and scholars within the movement itself.[37]

While not all of this activity emerged directly out of ecofeminism's originating tendencies, the contributions of the women involved in ecofeminism's early years are still very much felt today. The 'desire for nature' within radical feminism, social ecology, environmental justice, and international environmental politics gave rise to an ecofeminism that still expresses an embodied and non-hierarchical approach to the desire for nature that goes beyond individualistic and romantic tendencies within the wider ecology movement. Overall, ecofeminism has consistently offered a politicized and collective expression of a social, rather than individual, desire for political and ecological integrity. Striving to make connections between women's everyday lives and ecological degradation within the context of hierarchy and oppression, ecofeminism has continued to push the radical ecology movement forward by raising awareness of the ongoing need to examine issues of gender, culture, race, class, and power.

As we look toward the next decade, we may begin to consider how to continue to elaborate upon ecofeminist theory and action by building upon and transcending the possibilities and problems presented by its origins. By integrating new areas of ecofeminist scholarship with the best of what its 'originating traditions' have to offer, we may begin to explore the potentialities for creating an increasingly social 'desire for nature' that can take U.S. passionately and thoughtfully into the next century.

Notes

1. Anne Koedt, "Women in the Radical Movement," in *Radical Feminism*, eds. Anne Koedt, Ellen Levine, and Anita Rapone (New York: New York Times Books Co., 1973), p. 318.

2. The term 'cultural feminism' emerged during the '70s as a way to point to essentialist notions of sexual difference that surfaced within feminist discussions of sexuality, gender, and culture; notions that were embedded in new reconstructions of women's cultural practices including women's music festivals, newspapers, and medical clinics. For an indepth look at one of the earlier critiques of cultural feminism, written during the thick of the feminist sexuality debates, see Alice Echols, "The New Feminism of Yin and Yang," in *Powers of Desire: The Politics of Sexuality* eds. Ann Snitow, et al. (New York: Monthly Review Press, 1983), pp. 439-460. For a more comprehensive discussion also see Echols, *Daring to be Bad: Radical Feminism in America 1967-75, (Minneapolis: University of Minnesota Press, 1989)*.

3. Audre Lorde, "An Open Letter to Mary Daly," in *Sister Outsider: Essays and Speeches by Audre Lorde* (New York: The Crossing Press, 1984), p. 67.

4. See Gloria Anzaldua and Cherrie Moraga, Second Edition, *This Bridge Called My Back* (New York: Kitchen Table Press, 1983).

5. bell hooks, "Rethinking the Nature of Work," in *Feminist Theory: from margin to center,* (Boston: South End Press, 1984) p. 98.

6. Lorde, *Sister Outsider*, p. 70.

7. Audre Lorde, *The Cancer Journals* (San Francisco: Spinster's Ink., 1980).

8. "WITCH statement", in *Sisterhood is Powerful* ed. Robin Morgan (New York: Vintage Books, 1970), p. 539.

9. Ibid., p. 539.

10. Susan Griffin, *Woman and Nature: The Roaring Inside Her* (New York: Harper and Row, 1978).

11. Carolyn Merchant, *The Death of Nature: Women, Ecology and the Scientific Revolution* (San Francisco: Harper and Row, 1980).

12. Griffin, *Woman and Nature*, p. xv.

13. However, it is vital to note that the emergence of an ecological sensibility within the feminist body politics of the New Left did not negate or even necessarily inform radical feminism itself. Today, strains of radical feminism continue to evolve independent of an ecological focus or analysis. An ecological orientation was not endorsed by radical feminists

who maintained that it detracted from an agenda that primarily addresses women's immediate needs for bodily integrity and civil rights.

14. Unity Statement—Women's Pentagon Action, 1980, in Ynestra King, *What is Ecofeminism?* (New York: Ecofeminist Resources, 1990).

15. Ynestra King, "If I Can't Dance in Your Revolution, I'm Not Coming," in *Rocking the Ship of State: Toward a Feminist Peace Politics,* eds. Adrienne Harris and Ynestra King (Boulder: Westview Press, 1989), p. 282.

16. See Gwyn Kirk, "Our Greenham Common: Feminism and Nonviolence," in *Rocking the Ship of State: Toward a Feminist Peace Politics,* eds. Adrienne Harris and Ynestra King (Boulder: Westview Press, 1989), pp. 115-130.

17. For a sensitive and thorough discussion of WomanEarth, as well as an exploration of issues of race and class in ecofeminist politics in general, see Noël Sturgeon, *Ecofeminist Natures: Race, Gender, Feminist Theory and Political Action* (New York: Routledge, 1997).

18. Sturgeon, *Ecofeminist Natures,* p. 82.

19. Murray Bookchin, personal communication, June 11, 1998.

20. As ecofeminism has grown in popularity, there has been significant confusion regarding the origins of the term and of the movement itself. While during the early 1980s, the term (still largely unknown in many feminist circles) was most closely associated with the Women's Pentagon Action of which King was a primary organizer, the mid- to late-1980s brought newcomers unfamiliar with the movement's origins.

In recent years, many have attributed the origins of the term 'ecofeminism' to an article written in 1974 by Françoise d'Eaubonne entitled *Le Féminisme ou la Mort,* (Paris: Pierre Horay, 1974). However, the article did not reach English speaking audiences until 1994 (in an essay translated by Ruth Hottel as "The Time for Ecofeminism," in Carolyn Merchant, ed., *Key Concepts in Critical Theory: Ecology* (Atlantic Highlands, NJ: Humanities Press, 1994), almost fifteen years after the theory and movement had emerged as a way to explicitly link an anti-militarist, anti-capitalist, and anti-patriarchal stance to questions of ecology. Though a version of the d'Eaubonne essay did appear in 1980, in Elaine Marks and Isabelle de Courtivon, eds. *New French Feminisms: An Anthology* (Amherst: University of Massachusetts Press, 1980), this version does not explicitly mention ecofeminism.

Examining the lineage of the term is a way to explore the specific historical context in which ecofeminist theory and action emerged. Attempts to trace the ecofeminist movement itself back to d'Eaubonne obfuscate the historical continuity between ecofeminist curriculum and writing that emerged at the ISE by King, and the wider context of the U.S. New Left made up of activists involved in the radical feminist movement, the feminist peace movement, the anti-war movement, and the anti-nuclear movement.

21. Indeed, many of the anthropological texts written by feminists during the late 1960s and early '70s used the domestic/public split as a key analytical framework. For a glimpse into this discussion, see *Woman, Culture & Society,* eds. Michelle Zimbalist Rosaldo and Louise Lamphere (Stanford: Stanford University Press, 1974).

22. King continued to teach at the ISE through the 1980s and participated in the ISE's annual colloquium on ecofeminism until 1994. For a more comprehensive discussion of the relationship between King, Bookchin, and the ISE, see Noël Sturgeon's book *Ecofeminist Natures* (London: Routledge, 1997), pp. 32-40.

23. Ynestra King, "What is Ecofeminism?" in *What is Ecofeminism,* ed. Gwyn Kirk (New York: Ecofeminist Resources, 1990), p. 26.

24. Robert D. Bullard, Introduction in *Confronting Environmental Racism: Voices from the Grassroots,* ed. Robert D. Bullard (Boston: South End Press, 1993), p. 9.

25. Taylor, Dorceta E. "Environmentalism and the Politics of Inclusion." *Confronting Environmental Racism: Voices from the Grassroots,* ed. Robert D. Bullard (Boston: South End Press, 1993), p. 58.

26. Cynthia Hamilton, "Women, Home, and Community: The Struggle in an Urban Environment," in *Reweaving the World: The Emergence of Ecofeminism,* eds. Gloria Orenstein and Irene Diamond (San Francisco: Sierra Club Books, 1992), p. 217.

27. Gita Sen and Caren Grown, *Development, Crises, and Alternative Visions* (New York: Monthly Review Press, 1987), p. 29.

28. Vandana Shiva has contributed profoundly to a historical and anti-capitalist ecofeminist critique of the intersection between patriarchy, colonialism, global capital and ecological degradation. See Vandana Shiva and Maria Mies, *Ecofeminism* (London: Zed Books, 1993).

29. In 1987, I coined the term "social ecofeminism" to clarify a specifically leftist trajectory within a steadily differentiating ecofeminist milieu. That year, the term was embraced by the Left Green Network that included social ecofeminism as one of its "Ten Key Values".
In 1989, the Youth Greens embraced a social ecofeminism as well. Within these green forums and at the ISE, the term referred to an approach to ecofeminism informed by social anarchism and social ecology; it reflected an attempt to combine an historical understanding of questions of nature and gender with a reconstructive and utopian vision of a post-capitalist, post-statist society.

30. Judith Plant, "Introduction," in *Healing the Wounds: The Promise of Ecofeminism*, ed. Judith Plant (Philadelphia: New Society Publishers, 1989), pp. 1-7.

31. Many of the essays within *Reweaving The World* were originally presented as papers at the Ecofeminist Perspectives: Culture, Nature, Theory conference held at the University of Southern California in 1987.

32. In the early 1990s, there emerged a body of critical writings about the relationship between ecofeminism and questions of spiritualism, essentialism, and hegemony surrounding Third World development. See Ynestra King, "Ecofeminism: The Necessity of History & Mystery," in King, *What is Ecofeminism* (New York: Ecofeminist Resources, 1990). Also, for a more controversial discussion, see Janet Biehl, *Rethinking Ecofeminist Politics* (Montreal: Black Rose Books, 1991) and Catriona Sandilands, "Ecofeminism and It's Discontents: Notes Toward a Politics of Diversity," in *Trumpeter*, 8:2 Spring 1991. See also Cecile Jackson, "Women/Nature or Gender/History? A Critique of Ecofeminist Development," in *The Journal of Peasant Studies*, Vol. 20, No. 3. April 1993, pp. 389-419. Chris J. Cuomo also offers an interesting discussion of anti-essential criticism in *Feminism and Ecological Communities* (London: Routledge, 1998).

33. Charlene Spretnak, "Ecofeminism: Our Roots and Flowering," in *Reweaving the World: the Emergence of Ecofeminism*, eds. Irene Diamond and Gloria Feman Orenstein (San Francisco: Sierra Club Books, 1990), p. 6.

34. Ibid., p. 10.

35. See Greta Gaard, "Misunderstanding Ecofeminism," *Z Magazine* 3 (1) (1994): 22.

36. For a look at ecofeminists discussions of animal liberation that appeared in the early 1990s, see Greta Gaard's anthology *Ecofeminism: Women, Animals, Nature* (Philadelphia: Temple University Press, 1993).

37. See Greta Gaard, *Ecological Politics: Ecofeminists and the Greens* (Philadelphia: Temple University Press, 1988); Noël Sturgeon, *Ecofeminist Natures: Race, Gender, Feminist Theory and Political Action,* (London: Routledge, 1987); and Chris Cuomo, *Feminism and ecological communities: an ethic of flourishing* (London: Routledge, 1998).

Part II

THE NATURE OF DESIRE

CHAPTER THREE

THE NATURE OF SOCIAL DESIRE: SOCIAL ANARCHISM, FEMINISM, AND THE DESIRE TO BE SOCIAL

To create a truly radical approach to ecological politics, we must move discussions of ecology from the realm of romantic desire toward a new kind of social desire for a just and ecological society. This chapter represents a step toward that end by tracing developments within the West of a desire to be social in the broadest sense, a kind of sociality that highlights the potential for pleasure and meaning within a range of social and cooperative activities. By exploring the social, rather than individualistic or romantic side of desire, we may begin to understand our place within a wider social and ecological community, understanding in turn an expression of desire found in the history of social anarchism and in the new social movements that began in the sixties. This 'social desire,' or desire to be social, assumes a variety of forms ranging from a nascent anarchist impulse of centuries ago, to an explicitly social understanding of the erotic articulated within radical feminism of the seventies and eighties. By exploring desire from a social perspective, we may begin to appreciate new ways of constituting ourselves as subjects capable of creating the ecological society we so desire.

ANARCHISM: THE DESIRE TO BE SOCIAL

The anarchist tradition offers a rich and varied vision of an ethical and cooperative new world. While offering a range of often conflicting reconstructive visions, most anarchists share a value of mutualism and sensuality, portraying humanity as potentially cooperative and sociable. The most crude definition of modern anarchism is derived from the literal translation of the term, without rule, which reduces anarchism to a rejection of any kind of social, economic, or governmental organization. However, anarchism has a far more nuanced history that includes a variety of complex interpretations of exactly what without rule means. For instance, while many anarchists agree on the need to abolish the state, not all agree that all forms of governance should be abolished. In turn, while most anarchists agree that capitalism should be transcended, there exist a variety of interpretations regarding the role of production and labor in creating the new society. Questions regarding what kind of non-statist governance, or what kind of non-capitalist economic system to adopt, remain to be sorted out by anarchists today.

Beginning in the 13th century with the Brethren of the Free Spirit, through to the social anarchists of the nineteenth and twentieth centuries, and finally resurfacing in the counter-culture of the 1960s and 1970s, the anarchist impulse has continued to offer a vision of society based on a sensual and social understanding of the potentialities of human nature and desire. Like liberalism, the social tradition finds its roots within the womb of the old society, within the Middle Ages of Europe. But while liberalism was marked by a capitalist response to the breakdown of the feudal order, the early pre-anarchist and anarchist impulse represents a response that was overwhelmingly anti-capitalist. In turn, whereas most liberal theorists condoned the emergence of the nation-state within Europe and North America, many early anarchists opposed the formation of the state in general.

As early as the thirteenth century, Medieval socialists expressed a nascent anarchist impulse. During this time, there developed a series of popular sects ranging from religious and ascetic, to secular and hedonistic. One sect in particular, the Brethren of the Free Spirit, was marked by an undeniably pre-anarchist impulse. During the thirteenth and fourteenth centuries, the Brethren of the Free Spirit formed a loose confederation of sects in the Rhineland of central Germany.[1] Resisting institutions of class in general, the Brethren of the Free Spirit appeared primarily in towns marked by the struggle between the artisan class and the rising class of bourgeois patricians. The Free Spirit maintained that "a handmaiden or serf should sell their master's goods without his permission, and should refuse to pay tithes to the Church."[2]

Since the Brethren of the Free Spirit asserted that the Holy Spirit dwelled within each person, they advised that grace should be derived from the individual rather than from the Church. Promoting a hedonistic way of life, the Brethren of the Free Spirit encouraged the pleasures of sumptuous food, dress, and sexual promiscuity. Their emphasis on sensuality represents a striking departure from other similar pre-anarchist Medieval sects which merely promoted a kind of happiness derived from adherence to an ascetic life. The

Brethren of the Free Spirit, like many hedonistic sects of the time, had begun to explore the utopian and social dimensions of sensuality, articulating the relationship between ideas of freedom and desire. As Bookchin points out, the Free Spirit's "concept of freedom was expanded from a limited ideal of happiness based on the constraints of shared needs, into an ideal of pleasure based on the satisfaction of desire."[3]

Over the next several centuries more formal expressions of the anarchist impulse developed, articulated in less hedonistic terms. Nonetheless, the Brethren of the Free Spirit's desirous tendency, retained within many contemporary expressions of anarchism, linked the demand for desire to the demand for social freedom.

Social Anarchism: The Dialectic Of Desire And Structure

Although anarchism represents a varied and often misunderstood body of ideas, it is possible to point to a tendency within anarchist history, a *social anarchism*, that represents a challenge to classical liberal precepts of individualism and competition, proposing instead values of collectivity and cooperation. Social anarchism finds its origins in the works of such thinkers as Pierre Proudhon, Peter Kropotkin, Emma Goldman, Errico Malatesta, and the Spanish Anarchists as well as contemporary thinkers such as Murray Bookchin. Unlike their liberal counterparts, these thinkers propose a cooperative vision of society infused with constructive expressions of social desire.

Social anarchists unsettle the classical liberal assumption that human nature is primarily individualistic and competitive. According to classical liberal theorists such as Locke and Mill, the individual exists prior to society, and society represents a social contract between abstract individuals whose primary wish is to protect their own self-interest. In contrast, for social anarchism, society emerges out of both the material need for interdependence in addition to the desire to be social. Assuming the social group before the individual, social anarchism predicates the viability and pleasure of the individual upon that of the social group as a whole. Social anarchism recognizes the potential of individuals to mediate, rather than negate, their desire in a way that reflects responsibility to the larger group. Individual desire is both informed by the social group while also informing that group, allowing individuality and sociality to be constitutive of a social whole.[4] For Italian anarchist Errico Malatesta, the human spirit is characterized by an implicit desire to be social: a desire embedded in a matrix of symbolism, meaning, and self-sacrifice, that cannot be reduced to material necessity.

> ...this need of a social life, of an exchange of thoughts and feelings, has become for [human beings] a way of being which is essential to our way of life and has been transformed into sympathy, friendship,

love and goes on independently of the material advantages that association provides, so much so that in order to satisfy it one often faces all kinds of suffering and even death.[5]

Unlike classical liberal theory that portrays *competition* for material advantage as a primary human motivation, social anarchism identifies competition as but one human proclivity nurtured by hierarchical structures themselves. Further, social anarchism embraces a dialectical understanding of the complementary relationship between individuals and society.

For social anarchists, it is not 'human nature' in general, but hierarchy in particular, that inhibits the potential for true social maturity. It is social hierarchy that facilitates the emergence and perpetuation of anti-social behaviors such as greed, competition, alienation, and violence. In this way, social anarchism is not only a philosophy of human 'nature'; it is also a philosophy of social structure. Ironically, social anarchists, parodied as 'lovers of chaos', have often been extremely attentive to structure, for they realize that particular forms of structure either inhibit or nurture positive human potential for cooperation and sociality. For Goldman, the challenge for social anarchists is to create structures that are free of 'rule over', authority or hierarchy; to create structures that will restore to humanity the possibility for mature and liberatory association:

...government, with its unjust, arbitrary, repressive measures, must be done away with. Anarchism proposes to rescue the self-respect and independence of the individual from all restraint and invasion by authority. Only in freedom can [human beings] grow to their full stature. Only in freedom will [we] learn to think and move, and give the very best of ourselves. Only in freedom will we realize the true force of the social bonds which knit us together, and which are the true foundation of a normal social life.[6]

Indeed, social anarchists do not embrace a naively optimistic view of human nature. In fact, they often maintain a keen and sober understanding of the potential for individuals to abuse power when placed in positions of authority. If social anarchists are optimistic about anything, it is about the potential to create modes of social organization that bring out the very best in humanity. For social anarchism, it is not that people are always good or altruistic. Rather, social anarchism appreciates the fact that centralized and hierarchical structures allow those who are anti-social to make everyone else's lives miserable.

Desire and structure, then, work together dialectically so that the creation of socially desirable structures allows for the constructive expression of desire. It is out of social empathy and rationality, impulses that cultivate a movement toward the joy and freedom of the collective, that social anarchists create structures that allow the most freedom and expression to the widest number of people. And from the anarchist emphasis on structure flows an attention to the quality of the means as well as the end. While more authoritarian theories, such as Marxism, posit the state as a transitional and necessary structure, social anarchists do not tolerate expressions of hierarchy at any point in the revolutionary process. For social anarchism, the revolution itself represents an educational process that transforms each individual into the kind of person desirable for the new society. In order for this gradual transformation to take place, the process of revolution must embody the same values and structure of the good society itself.

Social anarchism focuses both on improving the quantitative material aspects of life and on improving the qualitative, sensual aspects of life. Expanding the revolutionary vista to include demands for roses as well as bread, social anarchism emphasizes the desire for beauty, pleasure, and self-expression in addition to emphasizing the desire for economic abundance and social cooperation. For Goldman, the process by which we transform society must be infused with degrees of meaning, sensuality, and pleasure that will characterize the new society we struggle to create. Her often quoted statement, "If I can't dance in your revolution, I'm not coming," stands as an emblem of the social anarchist appreciation for the crucial role that desire plays in the political struggle.[7]

New Left Desire: Social Desire Within The New Social Movements

Goldman's appeal for a revolution that makes room for dancing, a revolution that answers to the call of desire as well as need, was largely overshadowed by the Marx-influenced movements of the Old Left. For Marx, it is through material production that society achieves freedom from conditions of universal scarcity or need. Thus, it identified social relations of production as the primary focus for revolutionary activity.

With the emergence of the New Left, however, we see a revival of an old anarchist sensibility: a proclivity to widen the political agenda beyond ideas of need and social relations of production to re-encounter understandings of desire and social relations in general. In the United States, the civil rights, anti-war, student, and women's movements, demonstrated a new political sensibility that stretched the productionist perimeters of the Old Left. Critiquing racial and sexual inequality, U.S. military aggression, and the rationalization of consumer capitalism, a new culture cried out against such institutions as racism, the government, the military, the university, the church, and the nuclear family. The dual appeals to anti-authoritarianism and desire constituted a qualitative sensibility that gave the New Left its anarchic flavor. Disenchanted with the current social order, American youth demanded a quality of life that was sensually engaging. By the early sixties, the movement had transformed the landscape of the Old Left, creating a new sensibility that resonated with that of the Brethren of the Free Spirit from centuries before.

The anarchist sensibility of the American New Left re-articulated the concept of social desire as an expression of desire informed by a social and political vision. While the civil rights movement called for an end to racial inequality, it also made pleas for universal 'brotherly love' and compassion; while the anti-war movement called for an end to military aggression, it also appealed to ideas of sexual and sensual liberation, painting placards with the slogan, 'make love not war'. The qualitative flavor of these events, emphasizing the quality of social relationships and artistic and sensual expression, represented a rejection of a society that had been eviscerated by a post-war era of gross commodification and social conformity.

The civil rights movement, whose ideals are most equated with the brilliant speeches of Martin Luther King, were also articulated within the literature of essayist and novelist James Baldwin. While Baldwin, as an African American gay man, addressed the need to overcome the material injustices of racism, sexism, and classism, he also wrote prolifically of the vital role that creativity and sensuality play in the struggle for society to reclaim its humanity. Like others of the New Left, Baldwin was critical of the qualitative impoverishment that characterized Anglo-American culture, an impoverishment that led many white Americans to appropriate the cultural riches of African American culture without questioning racial injustice. For Baldwin, the struggle to overcome cultural and social impoverishment intensified by racism entailed a qualitative reconfiguration of the psychic world itself. To overcome racism, Baldwin reasoned, white Americans must transform not only structural, but aesthetic and psychic practices, addressing deeper cultural and sensual longings:

> [Racial] tensions are rooted in the very same depths as those from which love springs, or murder. The white man's unadmitted, and apparently, to him, unspeakable—private fears and longings are projected onto the Negro. The only way he can be released from the Negro's tyrannical power over him is to consent, in effect, to become black himself, to become part of that suffering and dancing country that he now watches wistfully from the heights of his lonely power and, armed with spiritual traveler's checks, visits surreptitiously after dark.[8]

In the literary works of Baldwin we witness a valorization of social desire: an acknowledgment of the transformative role that desire, art, and empathy may play in remaking society itself. For Baldwin, the role of the artist is to "illuminate that darkness, blaze roads through that vast forest, so that we will not, in all our doing, lose sight of its purpose, which is, after all, to make the world a more human dwelling place."[9] Baldwin displayed a unique ability to seamlessly integrate themes of creativity and empathy within the political project, elaborating a new sensual-political sensibility that was to unfold throughout the course of the decade.

A focus on the qualitative dimensions of social transformation can also be found in the writings of Murray Bookchin. Whereas an explicit anarchism in the U.S. had been eclipsed by socialist movements of the pre-war period, anarchist thought was reintroduced in Bookchin's canonical work, *Post-Scarcity Anarchism*.[10] Written during the late sixties, the essays in *Post-Scarcity* heralded the potential for what Bookchin called a "social libido," a radical integration of reason and passion that he hoped would be fulfilled within the new social movements. While Bookchin emphasized the need to overcome material necessity, he also asserted the importance of expanding the revolutionary horizon to encompass qualitative concerns as well:

...the revolution can no longer be imprisoned in the realm of Need. It can no longer be satisfied merely with the prose of political economy. The task of the Marxian critique has been completed and must be transcended. The subject has entered the revolutionary project with entirely new demands for experience, for re-integration, for fulfillment, for the *merveilleux [marvelous].*[11]

What we see in Bookchin's early writings is an attention to the qualitative and subjective dimensions of the revolution, dimensions that could not be accounted for by Marxist-based theories that dissolved the individual into essentialist categories of history or society. As Bookchin states so passionately, "A revolution that fails to achieve a liberation of daily life is counterrevolution. The self must always be identifiable in the revolution, not overwhelmed by it."[12] For Bookchin, questions of desire and need constitute a complementary matrix through which to reconstruct society as a whole: while countering the fabricated scarcity of the post-war period by constructing social and political counter-institutions (institutions and practices such as decentralized participatory democracy, municipal economics, and ecological technologies), revolutionaries must infuse these new cooperative and decentralized structures with creativity and sensuality—a vitality that he recognized within the "social libido" of the new social movements.

During the same period, in Europe, a similar sensibility emerged, culminating in the May 1968 revolt in Paris. In 1957, inspired by earlier aesthetic movements such as the Symbolist, Dadaist, and Surrealist movements, a group of avant-garde artists and writers from across Europe formed the Situationists International (SI). While situationist writer Guy Debord called for an end to the passive spectatorship of consumer capitalism, Raoul Vaneigem, joining the SI in 1962, called for a "revolution of everyday life."[13] While retaining a Marxian emphasis on production (promoting a program of workers councils), the SI departed from Marx by broadening the revolutionary focus to include a wide range of qualitative, aesthetic, and sexual demands. Articles published in *Internationale Situationiste* convey the spectrum of political and cultural concerns, ranging from questions of urban planning (referred to as 'urban geography'); artistic intervention (which included public poetry writing and graffitti); critiques of cinema and language; political responses to the Vietnam and Algerian wars; and the situations in China and the Middle East.[14]

Distinctive of the SI was the ability to infuse an urban idiom of political reconstruction with a poetic idiom of everyday life. In a communiqué delivered during the 1968 occupation of the Sorbonne, the SI's "Occupation Committee of the Autonomous and Popular Sorbonne University" advised others to disseminate slogans by:

> ...leaflets, announcements over microphones, comic strips, songs, graffiti, balloons on paintings in the Sorbonne, announcements in theaters during films or while disrupting them, balloons on subway billboards, before making love, after making love, in elevators, each time you raise your glass in a bar.[15]

The committee's list of slogans, including "occupy the factories," "down with the spectacle-commodity society," "abolish alienation," and "humanity won't be happy until the last bureaucrat is hung with the guts of the last capitalist," reflects an analysis grounded in a set of cultural, political, and economic concerns.[16] The SI called for ordinary people to "construct situations" within urban centers to awaken others from the deep sleep of capital and state-induced passivity. In this spirit, they called for the construction of aesthetic and political activities such as street theater, poetry, and graffiti, as well as public 'play' or 'games'. Unsettling vernacular distinctions between actor and audience, spectators and spectacle, so integral to consumer society, the SI promoted a sensual and creative re-activation of a desire that had been blunted by life in a bureaucratic and capitalist society, a desire that would engender a new political and social reality:

> The really experimental direction of situationist activity consists in setting up, on the basis of more or less clearly recognized desires, a temporary field of activity favorable to these desires. This alone can lead to the further clarification of these primitive desires, and to the confused emergence of new desires whose material roots will be precisely the *new reality* engendered by the situationist constructions.[17]

However, despite these promising expressions of social desire, by the end of the decade the social potentiality of this renewed desire remained unfulfilled. Between the early sixties and the Woodstock years, crucial tensions inherent within the new social movements in both the U.S. and in Europe came to the surface. Tensions between 'individualism' and 'individuality', tensions between an ardent egoism and a sense of selfhood grounded in a wider social consciousness and commitment emerged, making the movement susceptible to commercial appropriation. There is, indeed, always a tendency for social desire to 'break off' from the social and political project, expressing itself through cultural practices that emphasize individual satisfaction over the political project to liberate society as a whole. The tendency toward 'me-ism', so endemic to liberal capitalism in general, makes any qualitatively oriented social movement potential grist for the capitalist mill: the potential desire for social and political opposition is too often corralled into the desire for pseudo-oppositional fashion, music, and other expressions of life-style.[18]

In the case of anarchism, this tension may be attributed, in part, to an historical and unresolved relationship to the social contract theory of such classical liberals as John Locke and John Stuart Mill, and to the individualistic existentialism of Nietzsche. While *social* anarchism emphasizes the idea of an individual dependent upon and constitutive of a social whole, there exists among some anarchists, a liberal and existential tendency to view the individual as prior to, or independent of, the collective. And paradoxically, the individualist tendency within anarchism resonates with the liberal capitalist subject: an individual committed to promoting its own self-interest and pleasure. Hence, challenging the Marxian emphasis on production and need, an anarchist impulse surfaced within the new social movements in the U.S. and Europe, giving rise to a renewed expression of social desire. However, as the decade wore on, the dialectic of need and desire was upstaged by the dialectic between individualism and cooperation, a dialectic that yielded finally to a grossly commodified Woodstockian counter-culture based on individualistic cultural indulgence.

Yet, while the new social movements of the sixties were unable to fully actualize their potential to sustain and elaborate a truly politicized expression of social desire, they did achieve some remarkable feats. Critical of modern post-war society, the new social movements offered an approach to qualitative questions that was quite progressive in nature. Instead of blaming 'humanity', or a failed consciousness for social and cultural malaise, figures like James Baldwin, Murray Bookchin, and groups like the SI identified problems of economic and political structure, *while* attending to qualitative themes of desire, creativity, and 'love'.

Finding their roots in the Old Left, the anti-war, civil rights, women's, and Situationist movements were able to circumvent degrees of abstraction and romanticization that undermine the potential of the radical ecology and mainstream environmental movements today. Indeed, the SI was not constrained by a backward-looking critique of modernity, a critique that juxtaposed an idealized rural past to an inherently flawed and fallen city. Instead, they proposed a reclamation of all that was liberatory within the modern city, a celebration of self-determination and poetry that could reinvigorate an urban life eviscerated not by 'modernity' and 'technology', but by state bureaucracy and capitalism. Bypassing a regressive anti-modernism, the new social movements offered a bold expression of social desire: a demand for new liberatory structures infused with sensuality and empathy rather than a sentimental plea to return to a pre-fallen world.

Feminist Eros: Second-Wave Desire

While the theme of desire was articulated within 'mixed' movements of the New Left, it was steadily being developed by an emerging radical feminist movement as well. Like social anarchism, feminists of the New Left demonstrated the need to transform the qualitative as well as the quantitative dimensions of society. Departing from their liberal feminist predecessors and contemporaries, a new breed of *radical* feminists sought more than just material and institutional justice and equality with men within the present society. In addition to justice, they demanded a free society in which women could create themselves anew on a qualitative level, innovating new forms of aesthetics, political organizing, theory, and sensuality.

As we have seen earlier, the second wave of feminism emerged within the context of the New Left, which at its inception was dominated by a needs-oriented approach to social change. And, while an anarchist dimension emerged within the New Left, there also flourished the influence of Marxism, Maoism, and other forms of socialist thought, yielding a rationalized, instrumental approach to politics that alienated many women within the movement. While leftist politicos fought for the satisfaction of material need, women were often told that the more qualitative changes that they sought were irrelevant to the "big work" of revolution.

Women grew critical of the contradictions between the New Left's values of equality and the hierarchical structures that characterized a majority of New Left organizations. The new student movement, positioning its materialist goals within the realm of necessity, often rationalized the deployment of hierarchical and authoritarian means to execute its plans for achieving justice. Here the relationship between needs and authority surfaced as the preoccupation with filling urgent 'necessary' needs led to an ends justifies the means approach to social change. As is often the case, a focus on 'necessary ends' tends to bring revolution into a more authoritarian mood as the goal of abolishing need is used to legitimize the implementation of authoritarian methods.

A women's liberation movement responded to the authoritarian and instrumental tendency in the New Left, uncovering a wider revolutionary project, one that integrated need with desire and ends with means. In addition to fighting for 'freedom from' economic oppression and male violence, women in the movement began to fight for a new articulation of desire. This new desire was framed as 'freedom to' pursue a range of sensual, creative, and political satisfactions, emerging from a sensitivity to the qualitative dimensions of social and political life.

THE Psychology Of DESIRE: TowaRd A SociAl ERos

While giving rise to a 'cultural feminism', radical feminism also ventured into such arenas as feminist sociology and psychoanalytic theory. By the late 1970s, feminist critiques of Freudian theories flourished, critiques that explored the implications of patriarchy for the construction of understandings of desire. Feminist sociologists and psychoanalytic theorists such as Nancy Chodorow and Jessica Benjamin were among the many whose writings had tremendous implications for a feminist reconstruction of desire.[14] In particular, these theorists examined the transformation of the qualitative dimensions of women's psychology, unsettling liberal and individualistic understandings of desire.

The search for a new understanding of desire reflects the quest for a qualitatively better way of being that these new theorists hoped would be more cooperative, non-hierarchical, and supportive of women's self-expression. Theorists explored the possibility of a feminist Eros, what I call a *socio-erotic*, a continuum of social and sensual desires endowed with ethical, personal, and political meaning. While traditionally the word 'desire' has had both sexual and social meaning, the word 'erotic' has maintained an exclusively sexual definition. By attributing a social meaning to the 'erotic', theorists translated understandings of satisfaction and pleasure into non-sexual realms such as work and friendship, endowing 'the erotic' with the vernacular qualities of everyday life.

In 1970, Shulamith Firestone articulated an understanding of the erotic that included a broader range of specifically *social* passions. In her groundbreaking book, *The Dialectic of Sex: The Case For Feminist Revolution*,[20] Firestone called for a wider demand for everyday pleasure, challenging "the concentration of sexuality into highly charged objects, signifying the displacement of other social/affection needs onto sex."[21] In a spirit akin with the Situationists, Firestone called for a re-invigoration of desire within an otherwise deadening everyday world:

Eroticism is exciting...life would be a drab routine without at least that spark. That's the point. Why has all joy and excitement been concentrated, driven into that one narrow, difficult-to-find alley of human experience, and all the rest laid to waste? There's plenty to go around [within] the spectrum of our lives.[22]

Soon, other feminists began to articulate the relationship between a narrow understanding of the erotic and an impoverished quality of everyday life within patriarchy. Critical of a process of socialization that teaches women to vicariously enjoy the pleasure of men and children, the movement demanded a broader range of social passions, both personal and political. Feminists began to expand the definition of the erotic, accommodating a new spectrum of sensual and social demands.

The feminist quest for a 'social desire' ran parallel to the critique of male defined desire and rationality as feminist theorists explored the psychological construction of the liberal male subject. Questioning ideas of male desire and behavior, theorists critiqued such institutions as 'romance', tying the concept to the problem of male domination in general. For Firestone, romantic desire constituted "a cultural tool of male power to keep women from knowing their condition, a cultural tool to reinforce sex-class," a form of "gallantry" that keeps women from recognizing their subordinated position "in the name of love."[23] In turn, feminist ethicists such as Carol Gilligan and Mary Belenky began to challenge the rationality of the liberal subject. For these thinkers, a male approach to epistemological questions, precludes ethical ways of knowing often characteristic of women and others marginalized from the public sphere of liberal capitalist society.[24]

As these theorists unraveled the male subject of liberal capitalist society, they uncovered a subject who possessed a rationality reduced to cool instrumentality, an individualism reduced to egoistic autonomy, and a competitive impulse coddled to the point of infantile aggression. Such a male subject, they reasoned, to function effectively within a repressive capitalistic society, required a dispassionate and unempathetic psychology: a detached posture conducive to a tolerance for competition.[25] Accordingly, feminists reasoned that it was women's marginalization from capitalist practice that allowed them to maintain degrees of 'relationality'. Within the female subject, these theorists uncovered a psychology more relational than autonomously egoistic, more empathetic than competitive: an understanding of selfhood derived from women's socialized role within the relational world of the home. Excluded from the realm of entrepreneurial competition, these theorists maintained, women had retained vital aspects of their humanity.

This discussion of women's relational orientation was accompanied by an exploration of expressions of 'relational desire'. In 1978, sociologist Nancy Chodorow wrote *The Reproduction of Mothering: Psychoanalysis and the Sociology of Gender*,[26] challenging the 'biological' origin of women's desire to mother. Exploring the social construction of a female 'relational self', Chodorow suggested that the same socialization process that led girls to want to become mothers also led girls to desire more 'relationality' in general. Suggesting an idea of a 'relational desire', Chodorow shed light on a desire distinct from a sexual desire for men, a desire for connection with women friends, sisters, and mothers. While historically women's desire had been primarily defined as either an irrational and carnal desire for men or as a self-less yearning to nurture children, Chodorow opened a window into a world where women desired other *women*, expressing a desire that could be constructive, relational, and social.

Chodorow was among the first to examine the very mechanisms by which women develop the desire to care for others, challenging the assumed primacy of the male figure in the formation of female desire. While Freud asserted that little girls invariably desire to bond with their fathers, Chodorow asserted that it is the mother that girls primarily desire: Whereas the mother is the primary caretaker during the early years of a child's life, she forms a primal bond and identification with her daughter; and it is from this bond that the mother becomes the prototype for women's lifelong relationships with other women. Thus, for Chodorow, while most little girls are socialized to become genitally heterosexual, they often maintain a strong and primal desire to bond socially with other women.

Feminist psychoanalytical theorist Jessica Benjamin also explored women's desire, unsettling the liberal portrayal of desire as inevitably individualistic and competitive. In her book *The Bonds of Love: Psychoanalysis, Feminism, and the Problem of Domination*,[27] Benjamin revealed a relational desire between a mother and her newborn. According to Benjamin, early child development can be seen as a dynamic development: a process potentially marked by increasing degrees of mutuality and cooperation between mother and child, a mutualism that may in turn lead to increasing levels of cooperation and greater selfhood for both. Displacing the idea of an 'innate' capitalist inclination for competition and hyper-individualism, Benjamin posited the possibility that we are born with the potential for social desire.

In pursuit of a social side of desire, Benjamin challenged the neo-Freudian theory of Margaret Mahler that portrays early child development as an inevitable conflict between mother and child; a conflict marked by a process of 'individuation' that entails that the child 'negate' its connection to its mother by separating from her. In contrast to Mahler, Benjamin proposed that the child actually develops in cooperation with the mother within a nurturing process of mutual recognition. In this way, Benjamin challenged the liberal, capitalist bias within Mahler's theory, a bias that privileged the idea of individual autonomy over the idea of a potentially cooperative and relational self. In Benjamin's view, individual development occurs within the context of a social desire for connectedness. In her studies of early child development, she documented moments of mutualism and cooperation between mother and child:

> Frame by frame analysis of mothers and babies interacting reveals the minute adaptation of each partner's facial and gestural response to the other: mutual influence. The mother addresses the baby with the coordinated action of her voice, face, and hands. The infant responds with his whole body, wriggling or alert, mouth agape or smiling broadly. Then they may begin a dance of interaction in which the partners are so attuned that they move together in unison.[28]

In this 'dance of interaction', Benjamin saw a way of relating untainted by inherent conflict between self and other. Moreover, for Benjamin, early experiences of mutual recognition "prefigure the dynamics of erotic life."[29] In sexual, erotic union, she maintained, we can experience that form of mutual recognition in which both partners lose themselves in each other without a loss of self, losing self-consciousness without loss of awareness.

Benjamin described a desire both to know and be known, a desire that is not only sexual, but is profoundly social and relational, a longing to become part of another while retaining individuality. This process of mutual recognition represents a 'socio erotic' dance of separateness and connection, a nuanced dialogue which actually enhances and develops the subjectivity of both dancers. Far from the liberal Freudian drama in which every self is assumed to desire either complete merging with or annihilation of the other self, Benjamin proposes a mutualistic and cooperative understanding of selfhood, a proposal that has revolutionary implications. Ultimately, Benjamin suggests a potential for a subjectivity that is socially prepared to be cooperative rather than biologically driven to compete; a subject equipped to engage in a socially and ecologically cooperative world.

However, while Chodorow and Benjamin challenged the biological argument for an 'inherent' competitive human nature and desire, their failure to fully historicize and politicize their argument limited the utopian potential of their conclusions. Using the white, middle-class, nuclear family as their subject, both Chodorow and Benjamin generalized from this subject to the rest of humanity. Indeed, both theorists insufficiently problematized the modern 'invention' of the nuclear family and were thus unable to adequately situate their study historically. Further, their proposals to create more cooperative and relational subjectivities did not sufficiently address the need for deep institutional change that extends beyond the nuclear family itself. Rather than challenge capitalist and state structures that nurture competitive and individualistic practices, the authors focused on retooling the parenting dynamics within the nuclear family.

Yet again, we may appreciate the emergence of an attempt to propose a new understanding of human nature and desire. Like the Situationists and social anarchists before them, these feminists looked beyond a reactionary 'returnist' outlook toward a reconstructive possibility of creating a new kind of subject able to cooperate and live harmoniously with others. Although neither theorist identifies as anarchist, both Chodorow and Benjamin expressed an implicitly anarchist challenge to the idea that hierarchy, hyper-individuation and domination are inherent, necessary, and universal. Rejecting romantic notions of selfhood, notions predicated on a self that finds love and security only through a dialectic of predation and protection, these theorists offer the possibility of a kind of sociality marked by mutualism, a desire to see the other as part of, yet excitingly distinct from, the self.

Toward A Socio-Erotic

Drawing inspiration from new psychoanalytic understandings of desire, other feminist theorists explored the radical potential of community, empathy, and a new way of being in the world. One of the most striking contributions of this new feminist culture was a new perspective on female sexual desire. The idea of female sexuality, framed historically as the realm of competition over men, of romance, and sexual domination, was now framed as the feminist desire to bond with other women, a desire to form mutualistic relationships poised on the intersection between autonomy and connection.[30] This new concept of woman-bonding acquired new meaning within the context of an emerging 'lesbian feminism' that captured the imagination of many feminists in the New Left, engendering new understandings of eroticism.

From the late sixties through to the early eighties, several feminists initiated discussions about a specifically 'lesbian' desire that was to be both sexual and social. In 1980, Adrienne Rich played a primary role in highlighting the social dimension of lesbian desire in her ground-breaking article, "Compulsory Heterosexuality and Lesbian Existence".[31] Drawing from Chodorow, Rich challenged the idea of women's innate desire for men. In this essay, Rich uncovered a continuum of non-sexual forms of bonding between women that have always existed within the context of patriarchy, despite the attempts of patriarchal institutions and practices to guarantee exclusive male access to women's attention and affection.[32]

Introducing the concept of the "lesbian continuum," Rich articulated a wide spectrum of social and sexual desires that women have expressed to each other throughout history. Rich encouraged feminists to expand the concept of 'lesbianism' to include a wider variety of relationships between women, including the sharing of a rich inner life, bonding against male tyranny, sharing of political support, resisting heterosexual marriage, and choosing, instead, female friendship:[33]

> As the term lesbian has been held to limiting, clinical associations in its patriarchal definition, female friendship and comradeship have been set apart from the erotic, thus limiting the erotic itself. But as we deepen and broaden the range of what we define as lesbian existence, as we delineate a lesbian continuum, we begin to discover the erotic in female terms: as that which is unconfined to any single part of the body or solely to the body itself.[34]

While Rich's concept of the lesbian continuum was highly controversial (accused by many of de-emphasizing the specificity of the oppression faced by women involved in same sex relationships), it constitutes a significant and historical attempt to recognize degrees of autonomy, intensity, and sociality within women's relationships; relationships that, according to Rich, have been consistently trivialized, discouraged, and obstructed throughout history. For Rich, women's desire to bond with, and care for, other women, is essential to the process of reconstructing society: Activities such as female friendship and mothering should be valorized for their potential to make social life more pleasurable, meaningful, and cooperative.

In 1978, Audre Lorde, feminist anti-racist activist, theorist, and poet articulated one of the most innovative and influential positions on women's social desire in her essay "Uses of the Erotic: The Erotic as Power".[35] In this landmark work, Lorde explored the erotic as a creative force, a way of knowing and being that becomes warped and distorted by racism, sexism, and other expressions of social hierarchy. For Lorde, the erotic constitutes a spectrum of social and sensual satisfactions ranging from the joy of engaging in passionate conversation to the pleasure of cooperative and meaningful work. In "Uses of the Erotic", Lorde was the first to explicitly develop a feminist 'erotic' that is social and sensual, endowed with revolutionary implications.

Audre Lorde's primary contribution to 'desirous discourse' was to explicitly broaden the definition of the erotic to include a spectrum of everyday practices. Unlike Freud, who examined the infusion of an often destructive sexual erotic into the realm of everyday life, Lorde highlighted the constructive potential of a social desire that could restore to everyday life dimensions of mutualism and creativity. And while Lorde did not identify as an anarchist, her concept of the erotic suggests an anarchist view of human nature, implying too, the utopian potentiality of desire. According to Lorde, "in order to perpetuate itself, every oppression must corrupt or distort those various sources of power within the culture of the oppressed that can provide energy for change."[36] The sources of power, then, to which Lorde refers, constitute an anarchist impulse, a proclivity toward non-hierarchy that is quashed by hierarchial systems of power. In this way, Lorde endows the erotic with an ethical dimension, establishing it as a quality of being against which all of our actions may be measured for ethical content and meaning. Lorde describes the erotic as an impulse that moves women to take creative and courageous action to fight racism and sexism to change the world. Lorde's erotic represents a *creative* and social force reminiscent of the "social bonds which knit us together" described by Emma Goldman nearly a half-century before.[37]

The revolutionary implications of Lorde's essay unfold as we follow its logic to its most reasonable conclusions: if we were to demand from our everyday lives the same pleasure and passion that we hope to find in sexuality, then we would have to make some pretty profound institutional changes. If such institutions as racism, sexism, capitalism, and the state make misery out of our work and political engagement, in turn making a misery out of our social, familial, and sexual relationships; if hierarchy and authority inhibit the cultivation of creativity, participation, and pleasure, then surely, fighting to restore the erotic means nothing short of a social and political revolution:

> For once we begin to feel deeply all the aspects of our lives, we begin to demand from ourselves and from our life-pursuits that they feel in accordance with that joy which we know ourselves to be capable of. Our erotic knowledge empowers us, becomes a lens through which we scrutinize all aspects of our existence, forcing us to evaluate those aspects honestly in terms of their relative meaning within our lives. And this is a grave responsibility, projected from within each of us, not to settle for the convenient, the shoddy, the conventionally expected, nor the merely safe.[38]

Lorde's essay conveys a desire to resist that which obstructs a free expression of creativity, political empowerment, and collectivity. It suggests that within all of us is a potential for a desire that is bigger than just sexual appetite. It is the appetite for efficacy in a world which de-skills us, a hunger for a kind of revolutionary competence. Lorde asserts that beneath layers of self-hatred, there often lies an untapped body of self-love and courage which could emerge into a revolutionary force so vast that it could transform not only women but the social and political landscape with its fierce intelligence. Hence, in *Uses of the Erotic: The Erotic as Power,* Lorde offers an invitation to women to demand pleasure, passion, and creativity in more aspects of their lives. By expanding the idea of desire, Lorde touched the wide range of social desires of many women.

Finally, feminist explorations of desire permeated a spectrum of literary genres. Indeed, both Lorde and Rich, women whose poetry, fiction, and theory enriched a radical feminist literary canon, were complemented by the works of other women committed to carving out new understandings of subjectivity and desire. In particular, this impulse found literary fulfillment within the fiction, theory, and poetry of Alice Walker, particularly within her novel, *The Color Purple*, published in 1982.[39]

In this story, Celie, a young African American woman comes of age, discovering within herself an erotic impulse, both sensual and revolutionary. Within the course of the novel, Celie falls in love with "Shug Avery," a sensual and spiritual mentor, who helps Celie to recognize her own intelligence, talent, and capacity for love. It is within the matrix of the relationship between these two women that Celie comes to experience Benjamin's mutual recognition: the experience of being recognized fully while recognizing the other. After a life of subjugation by men, Celie rises to claim her own power as the forces of sensuality, mutualism, and autonomy come together, bringing her to a state of self-love. Separating from ideologies of racism, sexism, and Christianity, Celie is finally free to see "the color purple," a metaphor for the new erotic Shug teaches Celie to recognize within her own body and in the rest of the natural world.

In the character of Shug Avery, Walker articulates a new understanding of the erotic that has anarchistic implications. No longer a stingy authoritarian creator, Shug's 'god' becomes a fecund, non-hierarchical and creative natural process to be enjoyed through sensuality, sexuality, and pleasure. In one passage, Shug explains to Celie the potential for complementarity between sensuality and ethics saying, "Oh God love all them [sexual] feelings. That's some of the best stuff God did. And when you know God loves 'em you enjoys 'em a lot more. You can just relax, go with everything that's going, and praise God by liking what you like."[40]

In *The Color Purple* , Walker displaces the romantic dialectic of predation and protection that characterizes most love relationships in literature. In her offer of love, Shug makes no pretense of 'protection'. Rather, she assists Celie as she faces the realities of her own oppression, encouraging Celie to claim her own freedom. In turn, Shug is no romantic hero 'gallantly' constraining her desire for Celie. Instead, she proudly offers to Celie her own sexuality in an ethics of 'impurity'. In this way, Shug celebrates her own body and the natural world appealing to a sexual ethics reminiscent of the Brethren of the Free Spirit. Walker conveys the possibility of a love between women that is neither idealized nor constrained, but delicious in its imperfection. While Celie adores Shug, she is able to recognize and accept Shug's weaknesses and failings. Walker transcends a liberal as well as romantic portrayal of desire, depicting a love that is unegoistic, a desire that seeks neither status nor triumph in 'winning'. In fact, Walker's depiction of Shug's non-monogamy illustrates a decidedly unproprietarian approach to love. Shug loves Celie in a spirit of mutualism, wanting only to further empower her to develop her own autonomy and potential for self-love, mutualism, and pleasure.

In *The Color Purple*, Walker paints a world that is both social and sensual, ethical and anarchistic. The life which Walker creates for Celie toward the end of the novel represents a metaphor for social utopia: a grand reconciliation of differences between the sexes and a reclamation of power, pleasure, and self-love by women. As we leave Celie, we find her living cooperatively within her small community of friends, engaged in work that she loves, generously giving to and receiving from her loved ones. Through the love of another woman, she has come home to herself, seated firmly at the center of her own ability to desire herself, others, and the natural world.

Hence, within second-wave feminism, we find a reach for a new "socio erotic," an understanding of desire that has distinctly social, and even revolutionary, implications. While Rich valorized the idea of women's mutualistic desire, Lorde elaborated a poetic and evocative exploration of a desire to reclaim a cooperative impulse in the face of such injustices as racism and sexism. In turn, in *The Color Purple*, we see a literary illustration of social desire: a story that explores the possibility for re-establishing new understandings of the impulse toward mutualism, interdependence, and sensual pleasure.

Perhaps most significant, we see in this 'erotic moment' a critique of modernity that is not regressive or romantic, but is decidedly forward looking. Critiquing such modern forms of hierarchy as racism, sexism, and capitalism, these theorists do not offer an anti-modernist alternative. Tracing hierarchies such as patriarchy back to pre-modern times, theorists such as Rich, Lorde, and Walker do not romanticize the past, blaming modern 'technology', 'urban life', or 'humanity' in general for causing social suffering. Instead, these theorists ground their critique in a historicized objection to practices of sexism and racism, offering possibilities for new forms of subjectivity that may emerge when people come to resist and transform these structures.

Further, the 'erotic' that these writers appeal to is not 'pre-modern', 'rural', or 'free' from humanity: That Celie faces the racist and sexist horrors of her childhood in a rural setting speaks to Walker's rejection of a romantic impulse that ignores a legacy of racism that still flourishes within the rural South as well as throughout the country.

Most important, in these critiques and reconstructions of modern desire, we see a utopian impulse that recognizes within the human spirit a potentiality for cooperation and ecological harmony: From the joyous mutualism depicted by Rich and Lorde to the 'ecological sensuality' depicted by Walker in the character of Shug, we see an antidote to the anti-humanism that marks much contemporary ecological discourse. Here we see an expression of the desire to be deeply related both socially and ecologically; a desire obstructed not by 'modernity' or 'humanity', but by social hierarchy itself.

Notes

1. See Jeffrey B. Russell, "The Brethren of the Free Spirit," in *Religious Dissent in the Middle Ages* ed. J.B. Russell (New York: John Wiley & Sons, 1971), p. 87-90.

2. Quoted in Murray Bookchin, *The Ecology of Freedom* (Palo Alto: Cheshire Books, 1982).

3. Ibid., p. 211.

4. Concerning questions of desire, the social tradition departs from the romantic and liberal traditions dramatically. If the romantic idealizes the exceptional qualities of a particular individual, the social anarchist recognizes the potential for exceptional qualities within the many. For those in the social tradition, the best in human nature is to be expected and encouraged by and for everyone, rather than being located within one ideal individual.

5. Errico Malatesta, *Anarchy* (Great Britain: Freedom Press, 1974), p. 26.

6. Emma Goldman, "Anarchism: What it Really Stands For," in *Anarchism and Other Essays* (New York: Dover Publications, 1969), p. 61.

7. In contrast to Freud, most social anarchists regard desire as a vital catalyst toward releasing the human potential for cooperation and dynamic self-governance within society. Social anarchism carries an implicit philosophy of desire, proposing that individuals can potentially express a wide variety of social desires when organized within 'desirable' non-hierarchical structures. For instance, Emma Goldman in her essay, "Sex, The Great Element for Creative Work," challenges the Freudian notion that creativity is made possible by the repression of sexual desire. She writes, "the creative spirit is not an antidote to the sex instinct, but a part of its forceful expression...Sex is the source of life... Since love is an art, sex love is likewise an art." In this way, Goldman maintained that sexual desire is not only compatible with, but actually complementary to, a full social life. See Candace Falk, *Love, Anarchy, and Emma Goldman* (New Brunswick: Rutgers University Press, 1984), p. 99.

8. James Baldwin, "The Fire Next Time," In *The Price of the Ticket: Collected Nonfiction 1948-1985* (New York: St. Martin's, 1985), p. 375.

9. Ibid., p. 315.

10. In Bookchin's *Post-Scarcity*, we see the emergence of an appreciation of the subjective dimensions of revolution that could not be accounted for by Marxist based theories. See Murray Bookchin, *Post-Scarcity Anarchism* (Montreal: Black Rose Books, Reprinted 1986).

11. Ibid., p. 307.

12. Ibid., p. 66.

13. Vaneigem's text, with the writings of Guy Debord, constituted a small but influential literary canon most associated with Situationism and the events of 1968. See Raoul Vaneigem, *The Revolution of Everyday Life*, trans. Donald Nicholson-Smith (London: Aldgate Press 1983).

14. For an exciting and well written discussion of Situationist history and implications for contemporary postmodern discourse, see Sadie Plant, *The Most Radical Gesture: The Situationist International in a Postmodern Age* (London: Routledge, 1992).

15. Quoted in *Situationist International Anthology*, trans., ed. Ken Knabb (Berkeley: The Bureau of Public Secrets, 1989), p. 344.

16. Ibid., p. 344.

17. Ibid., p. 43.

18. Decades after publishing *Post-Scarcity Anarchism*, Bookchin has come to reconsider his earlier enthusiasm regarding the potential of a post-war generation to locate questions of subjectivity within a truly oppositional and revolutionary trajectory. While dismayed by the failures of the new social movements to transcend commercial cooptation, nihilism, and an egoistic 'me-ism', Bookchin sees in much of today's expressions of anarchism a continuation of this disappointing trend. For a provocative discussion of such issues, see Bookchin, *Social Anarchism or Lifestyle Anarchism: An Unbridgeable Chasm* (London: AK Press, 1995).

19. Beginning in the seventies, a school of feminist psychology emerged in dialogue with a range of feminist epistemologists, ethicists, sociologists, and feminist historians of science. Reconsidering discourses such as modern science and psychoanalytic theory, feminists challenged notions of universal objectivity, rationality, and competition, offering insights into the 'relational' subjectivity of women and other marginalized peoples. The reconstructive vision that emerged from these forums focused primarily on re-orderings of social and cultural institutions of family, education, and scientific production. See Jean Baker Miller, *Toward a New Psychology of Women* (Boston: Beacon Press, 1976), Dorothy Dinnerstein, *The Mermaid and the Minotaur : Sexual Arrangements and Human Malaise* (New York: Harper Colophon Books, 1976), and *Women's Ways of Knowing: The Development of Self, Voice, and Mind*, eds. Mary Field Belenky et al. (New York: Basic Book Publishers, 1986). Also see Carol Gilligan, *In a Different Voice.: Psychological Theory and Women's Development* (Cambridge: Harvard University Press, 1982), Catherine Keller, *From a Broken Web: Separation, Sexism, and Self.* (Boston: Beacon Press, 1986). Also, two particularly good anthologies to emerge from these discussions are *Gender/Body/Knowledge: Feminist Reconstructions of Being and Knowing*, eds. Alison M. Jaggar and Susan R. Bordo (New Brunswick: Rutgers University Press, 1989) and *Women's Consciousness, Women's Conscience*, eds. Barbara Hilkert Andolsen et al. (San Francisco: Harper and Row, 1985). Both Evelyn Fox Keller and Donna J. Haraway have contributed significantly to a new feminist approach to questions of scientific objectivity and knowledge production in general. See Evelyn Fox Keller, *Reflections on Gender and Science* (New Haven: Yale University Press, 1985), Donna J. Haraway, *Simians, Cyborgs, and Women: The Reinvention of Nature* (New York: Routledge, 1991).

20. Shulamith Firestone, *The Dialectic of Sex: The Case For Feminist Revolution* (New York: Bantam Books, 1970).

21. Ibid., p. 67.

22. Ibid., p. 159.

23. Ibid., p. 147.

24. See Carol Gilligan, *In a Different Voice: Psychological Theory and Women's Development* (Cambridge: Harvard University Press, 1982), and *Women's Ways of Knowing: The Development of Self, Voice, and Mind*, eds. Mary Field Belenky et al. (New York: Basic Book Publishers, 1986).

25. For a feminist discussion of 'competition', see *Competition: A Feminist Taboo?* eds. Valerie Miner and Helen E. Longino (New York: The Feminist Press, 1987).

26. Nancy Chodorow, *The Reproduction of Mothering: Psychoanalysis and the Sociology of Gender* (Berkeley: University of California Press, 1978).

27. Jessica Benjamin, *The Bonds of Love: Psychoanalysis, Feminism, and the Problem of Domination* (New York: Pantheon Books, 1988).

28. Ibid., p. 155.

29. Ibid., p. 147.

30. Ann Snitow offers an intriguing, yet controversial discussion of the political context surrounding lesbian feminism in the wider feminist movement. According to Snitow, lesbian feminists broadened the concept of lesbian desire beyond sexuality for a few reasons. First, she contends, lesbians sought to build acceptance within a larger, historically heterosexist feminist movement. As a way to build bridges with heterosexual women in the movement, she maintains, lesbian feminists defined lesbianism as but one expression of desire between women, thus situating lesbianism within the scope of a greater 'sisterhood.' For Snitow, this attempt was part of an even larger feminist project to reconstruct not only desire but society as a whole on feminist terms. Second, according to Snitow, lesbian feminists often de-emphasized the sexual aspect of lesbian desire in order to differentiate lesbian feminism

from male defined lesbian images portrayed in mainstream heterosexual pornography which present lesbian identity in male terms. See Alice Echols, *Daring to be Bad: Radical Feminism 1967-1975* (Minnesota: University: 1989).

31. Adrienne Rich, "Compulsory Heterosexuality and Lesbian Existence," in *Powers of Desire: The Politics of Sexuality*, eds. Ann Snitow et al. (New York: Monthly Review Press, 1983).

32. Ibid., p. 177-202.

33. Ibid., p. 192.

34. Ibid., p. 193.

35. Audre Lorde, *Sister Outsider* (New York: The Crossing Press, 1984).

36. Ibid., p. 55.

37. Emma Goldman, *Anarchism and Other Essays*, p. 64

38. Audre Lorde, *Sister Outsider*, p. 53.

39. Alice Walker, *The Color Purple* (New York: Pocket Books, 1982).

40. Ibid., p. 203.

CHAPTER FOUR

THE FIVE FINGERS OF SOCIAL DESIRE:
THE DIMENSIONS OF THE SOCIO-EROTIC

The cultural landscape within the age of global capital leaves much to be desired. Looking out across any small town, suburb, or city in the United States we can detect two yellow glints: McDonald's arches poking up into the sky, competing with the white church steeples that used to dominate the horizon. The glaring signs of fast-food chains and the endless sound bites of telecommunications are tropes of a brave new service economy, an economy that has been equated with the de-spiritualization of culture itself. Capitalist standardization and regularization have encroached into our everyday lives, reducing social, cultural, and political relationships to 'consumer' and 'producer' as we buy and sell standardized food, infotainment, health care, new age religion, education, and even political representatives. In turn, as the cultural landscape succumbs to social alienation and erosion, the 'natural' landscape deteriorates as well. Each night, newscasters announce the arrival of yet another 'endangered species' or a 'disaster of the week', another hurricane, tornado, earthquake, or flood resulting from greenhouse-induced climatic instability. And while the natural world is literally disintegrating, it is also being rationalized on unprecedented levels—reduced to genetic 'natural resources' to be surveyed, patented, and sold for profit.

How we interpret these events is deeply significant. Whether we attribute these 'rationalizations' to a failed spiritual or romantic orientation or to centuries of capitalist driven industry and an authoritarian state, such interpretations have tremendous implications for how we address problems of social and ecological injustice. Whereas a focus on the former tends to bring the revolution into a more contemplative and individualistic mood, the latter opens the way for a critique of hierarchical institutional structures. Yet it is not necessary to engender a false dilemma between spirituality and politics in order to address issues of social and ecological change. Rather, we may develop new ways to talk about questions of meaning, quality, sensibility or spirituality, ways that are integral to talking about institutional and political change. For the common link between ideas of meaning and ideas of structure is the idea of relationality. The idea of social *relationships* is integral to the idea of social *structures*—non-hierarchical structures that facilitate *meaningful* cooperative social relationships in all areas of our lives.

This chapter initiates a discussion of how to re-cast common understandings of 'meaning' that are conventionally framed in spiritual or romantic terms, ways to discuss those meaningful aspects of social and ecological life that are degraded by capital-driven technology and state formations, ways to talk about those aspects of reality that cannot be reduced to capitalist rationalization with its productionist idiom of means-ends, 'bottom lines', or standardization. Moving beyond dualistic concepts such as 'spirit' provides the opportunity to cultivate new metaphors for articulating that which is intensely meaningful and connective, metaphors that are derived from a relational tradition of *Eros*. By shifting from discussions of spirituality or romantic idealization to idioms of the erotic and social desire, we are better able to transcend binaries between the spiritual and the political that currently limit discussions of social and ecological justice.

Beyond Rationalization: From Spiritus To Eros

The McDonaldsization of culture is often associated with the dramatic decline in the quality of social and ecological relationships. Reducing social relationships to predetermined interactions between server and servee, each aspect of a McDonald's is prescribed, regularized, number-crunched, and market-analyzed. The McDonald's idiom is so embedded in everyday cultural practice that McDonald's itself may serve as a symbol of the cultural effects of advanced capitalist rationalization.[1] McDonald's translation of assembly-line industrial practice to service production typifies all that is de-spirited within 'advanced' capitalism.

However, the problem of capitalist rationalization has a history that began long before the appearance of those plastic golden arches. At the turn of the century, Max Weber described the disenchantment of everyday life and work due to modern capitalist rationalization.[2] For Weber, a rationalized capitalism implied a disciplined labor force and the regularized investment of capital, practices that entail the continual accumulation of wealth for its own sake. Contemporary critiques of such principles as 'profit over quality of life', 'regularization over individual expression', and 'standardization of everyday life', are often derived from Weber's description of the cultural implications of a modern capitalism.

Yet Weber's crucial insights into the cultural implications of capitalism have often been upstaged by popular critiques of modernity that emphasize 'rationality' and 'spiritual decay' as causes of an impoverished quality of everyday life and work. As in the case of early eco-fascism in Germany, instead of critiquing capitalist rationalization, theorists blamed modern rationality for society's ills.[3] And rather than fight capitalism by creating cooperative social and political institutions, such critics fought the cultural and ecological *effects* of capitalism by proposing a spirituality and anti-rationality that would either co-exist with, or perhaps reform, the capitalist system.

Yet the cause of cultural and ecological degradation is indeed capitalist rationalization, not a modern fall from spiritual grace. And if capitalism is a set of *social* relationships based on exploitation, regularization, alienation, and commodification, then the antidote to capitalist rationalization is a new relationality, an empathetic, sensual, and rational way of relating that is deeply cooperative, pleasurable, and meaningful.

Instead of pitting the idea of spirit against the idea of rationality, we need to cultivate a new rational and empathetic orientation capable of de-stabilizing capitalist rationalization. We need to move beyond a focus on spirituality to a focus on a rational and empathetic relationality to create institutions that will nurture cooperative ways of relating socially and ecologically. However, the shift from spirituality to a relationality entails a great leap for Westerners steeped in normative dualisms between spirit and matter, or intuition and rationality. Just as we learn that black is the opposite of white, we learn that rationality is the opposite of intuition and spirituality. Accordingly, when disenchanted by a rationalized and 'McDonaldsian' world, we confuse rationalization with rationality, and look immediately to intuition and spirit for both solace and a solution.

Today, when we appeal to the term spirituality to discuss cultural and ecological meaning, we end up taking home more than we bargained for. Anchoring contemporary ideas of social and ecological integrity to ancient dualistic 'activating principles' perpetuates reductive and polarized understandings of reality. The term spirit is embedded within the psychic trenches of Western metaphysical dualism. Its origin can be traced to the Latin *'spiritus'*, an 'activating principle' that was believed to animate an inert, feminine, and passive body with the invigorating properties of breath. According to the ancient Romans, it is when we breathe (*spirare*) an eternal breath (*spiritus*) that an otherwise inactive and ephemeral body comes to life. Conversely, it is when spiritus leaves the body that we die.

And when we blend this Western notion of spirituality with non-Western systems of meaning, we face another set of problems. The journey from a non-Western language into the language of spiritus is a tricky one indeed. Hopes to find in pagan, Neolithic, Eastern, and indigenous religious practices, a non-dualistic understanding of spirit are undermined by appeals to a dualistic linguistic tradition of spiritus; a tradition predicated on ideas of activating principles counterposed to a passive matter. While the idea of spiritus, or breath, is appealing to ecologically oriented theorists, for the ancient Romans, spiritus entailed a breath that activated an otherwise dead body. Today we know that breath does not activate, but rather, is functionally integral to a body that is already very much alive.

Still faced with the need for a metaphorical antidote to the problem of capitalist rationalization, a trend in society that cheapens all that is meaningful, we must engender other ways to articulate meaning. Disenchanted with capital-driven science and technics that promise to render all knowledge and experience 'operative', 'useful', and 'efficient', theologians are left with few alternatives (other than spiritus) for describing meaningful practice and perception. Such theorists yearn to be able to point to qualities of reality that are irreducible, qualities that cannot be known or conveyed through the language of logical positivism, behaviorism, biological determinism, or physics.[4] Moreover, such thinkers long to be able to convey the possibility of knowing the poetry of bodies and the natural world, illustrating the irreducible quality of the *connections* between bodies and within bodies themselves.

However, there is another tradition to which we may appeal. Leaving the world of spiritual metaphysics, we may engage another way of talking about meaning. There exists another kind of principle that, while not activating, or spiritual, is relational and social. The term 'Eros' contains an idea of love, an expression of desire between individuals. It is in the space between individuals, within the hearts of individuals, that Eros flourishes. Eros, then, represents an *embodied* quality of social relationships—an attraction, passion, and yearning of one self for other selves.

However, to emphasize the relational and social quality of Eros, we must first establish an understanding that is distinct from the Freudian definition that reduced Eros to a physical energy.[5] Freud reconstituted the idea of Eros into an energistic Life force that must be repressed in surrender to a civilizing reality principle. In the era of liberal capitalism, desire is often cast within energistic or individualized terms, and it is usually framed in terms of scarcity, as the will to overcome a particular deprivation, replacing desire with a particular object of want that is external to the self.[6] However, when we shake our theoretical kaleidoscope slightly, we may reconfigure the idea of desire as a will to express a potentiality that lies not outside of ourselves, but inside our very being, inside our social and political communities. We may articulate an idea of a potential to express sensuality, sociability, and creativity in all of its delectable complexity, a potential for social desire that exists within us at every given moment; not as an individual triumph over an inner emptiness, but as a social and cooperative expression of a fullness that yearns to emerge.

And yet, when we seek to elaborate discussions of social desire, we are confronted by a linguistic and conceptual vacuum: While the language of liberal capitalism offers a rich vocabulary for describing what is anti-social, it offers an impoverished vocabulary for describing the cooperative impulse. We know far more about anti-social, irrational desires such as greed, acquisitiveness, domination, and competition, than we do about desires that enhance the subjectivity of both self and other. In turn, as Michel Foucault points out, we are indeed saturated by discourses on 'sexuality'.[7] However, we have a paucity of discourses on social desires for creativity and solidarity.

As we move beyond an energistic Freudian idiom of forces, repression, drives, and release, Eros could represent a metaphor for sociality itself. The idea of Eros, or the more vernacular term, the erotic, provides a metaphor for a quality of social relationships that is passionate, loving, mutualistic, and empathetic. And building upon the idea of the erotic, we may point to a cooperative dimension of desire. We may speak of a socio-erotic, a spectrum of social and sensual desires that enhance social cooperation and a progressive revolutionary impulse.

The socio-erotic, as a metaphor for a relational orientation that may counter capitalist rationalization, places social and cultural criticism on much firmer ground. Instead of conflating rationalization with a rationality to be countered by an irrational spirit, we may appeal to the idea of a socio-erotic, a way of talking about an impulse toward collectivity, sensuality, and non-hierarchy that may be nourished and encouraged by the creation of non-hierarchical institutions. The idea of a socio-erotic, or a spectrum of social desires, is implicit within many feminist and social anarchist writings that reveal the delicate and crucial link between desire and freedom. The desire for a quality of life that is sensual, cooperative, creative, and ethical resonates with the impulse for a way of life that is not only based on justice and equality, but on a profound sense of freedom as well. The socio-erotic represents the spectrum of social desires that emerges from this longing for freedom, this impulse toward an interdependent and harmonious world. The very act of thinking through the socio-erotic represents an exercise in strolling the perimeters of a passionate landscape that could potentially encompass the full scope of our personal, social, and political lives.

The project to further elaborate understandings of desire is central to ecology. By exploring the social desire for ecological justice and integrity, we may begin to uncover new ways to articulate what it is that we really yearn for when we talk about 'nature'. Often framed in terms of a spiritual or romantic longing for connectedness, wholeness, and integrity, the social desire for nature is often contrasted to universalizing notions of rationality and technology that are accused of destroying all that is good in the world.[8] Again, conflating rationality with a particular kind of rationalization, 'nature lovers' often propose a return to an intuition and spirituality that would better resonate with ecological principles such as connectedness, diversity, or inter-dependence. However, as we shall see, it is possible to think rationally, with great feeling, about the social desire for nature. Instead of appealing to ideas of spirit and intuition to identify moments of meaning, connectedness, and integrity, we may appeal to the embodied and relational idiom of the socio-erotic.

Thε Fivε Fiηqεrs Of Social Dεsirε

When a child reaches out to the world, it reaches with both hands. Often, the child reaches for something it needs physically or for some form of social interaction that it desires. As we dive into the vast blue world of the socio-erotic, we no longer define desire as the singular will to satisfy an individualistic longing for that which we do not have, nor do we reduce desire to material need. Instead, we may explore desire as a rich dialectic, as a yearning to unfold all that we can feel and do together within a free society. In particular, social desire represents an organic and profoundly *social* spectrum of potentialities, inclinations, or tendencies. It represents a will to know ourselves, each other, and the world. From within this spectrum of social desire, there emerge five dimensions of desire, "five fingers of social desire," which are implicit within the social tradition itself. These dimensions are linked to the desire for sensuality, association, differentiation, development, and political opposition. And like the graceful movements of a hand, the socio-erotic can best grasp the world when all five fingers and palm work in unison.

Sεηsual Dεsirε: Thε Dεsirε To Kηow

Let us begin with one of the most common understandings of desire, one with which we are most familiar. The first finger of desire, *sensual desire*, is the desire for sensual expression, satisfaction, and engagement with any one, or all, of our senses. Sensual desire begins with the assertion, "*I want to know,*" sensually, engaging ourselves on a visceral level. The idea of sensual desire represents the most unmediated dimension of desire, referring to a will to know through the senses, to express our potential for sensual enjoyment and experience. When we think of sensual desire, we may think of the way children seek out the world through their mouths and fingers, yearning in return for nourishment and affection. We may let the little finger symbolize sensual desire, the desire to delight in our senses, which incorporates itself within all other dimensions of social desire.

Within sensual desire, we also immediately discover a dimension of social meaning, for we see that it is impossible to consider the idea of sensual desire without situating this desire within a specific social context. Indeed, there is no pre-social sensual desire. While infants are born with a suckling instinct, they must *learn* to respond to the world visually, tactually, and aurally. The ability to glean pleasure from gazing at the world, the ability to distinguish and interpret sensations around us emerges from the stimulation of caretakers who gaze into an infant's eyes, touching and cooing at them in an engaging manner. It is through being sensually stimulated within a social relationship, that infants develop the ability to recognize, integrate, and enjoy sensual stimulation. In this way, the capacity for cultivating and expressing sensual desire is predicated on a deeply relational social context.

In addition, sensual desire is culturally constrained. While we may desire sensual engagement through our senses by eating, drinking, hearing, smelling, or touching the world, the *way* in which we approach and encode these sensual practices is overwhelmingly informed by the culture in which we live. Similarly, the sensual desire for 'nature' is a social form of desire. In the West, for instance, from the day we are born, we develop culturally specific understandings of what we will categorize as 'natural' as well as what aspects of this 'nature' we will find appealing. As illustrated by theorist Donna Haraway, historical understandings of 'landscape', 'the pastoral', 'wilderness', and 'animality' inform the ability to identify and respond to those sensual aspects of ecological reality we take for granted as 'natural'.[9]

Sensual desire is contingent upon social, cultural, and political practices that establish the standards by which we distinguish such sensual values as beauty, strength, grace, and taste. Whether we express a desire to see, touch, smell, or talk to another person, this desire to associate sensually is both socially constrained and facilitated. And because we endow these social interactions with specifically sensual contexts, such as in the sharing of food, music, dance, or sexuality, we imbue these associative activities with a dimension of sensual desire as well.

Associative Desire: The Desire To Know Other

Associative desire, the second finger of social desire, adds another dimension by beginning with the assertion: "*I want to know you.*" Whereas association is not always explicitly 'physical' or 'sexual', there exists a dimension of sensuality within an association between people who feel related or bonded. This sensuality may range from the flow of voices or hand gestures of spoken communication, to the visual gaze between two people standing at opposite ends of a room. In turn, we may express our desire for sensual association through activities ranging from the breaking of bread to the sharing of sexual intimacy. Hence, we may allow the 'ring finger' to symbolize associative desire, representing the finger that is most associated with relationships, friendship, and love.

As we think through the dialectic of social desire, we must regard the metaphor of the hand as only a point of departure, asking our minds to do that which the static symbol of the hand cannot: our minds can think dialectically, allowing each dimension of social desire to be incorporated and integrated into the next, bringing a cumulative and non-linear fullness to our understanding of social desire. We may derive the idea of sensual desire from the idea of associative desire, allowing the one to give richness and meaning to the other. Hence, from the idea of sensuality, we may educe an idea of associative desire, mediating the idea of sensuality with the idea of association. Sensual, associative desire is what we commonly call 'love'; it is the expression of bonds of friendship or lovership, the desire to create and maintain bonds with family, community, and with the stranger for whom we feel empathy. While we may not always express overt sensual desire to those with whom we feel a connection, the very idea of 'feeling' a 'connection' conveys the ever present dimension of sensual desire within the associative moment.

Social anarchists ranging from Peter Kropotkin to Murray Bookchin have explored this desire for association, demonstrating its salience within the revolutionary project. Human nature is marked by tendencies toward both the social and the anti-social. It is however, the *social* tendency that represents the potential to be cooperative, to exist within a vital social matrix on which all depend. Associative desire acts as a glue which binds people together, allowing them to express the yearning to enhance the richness of each other's material and social lives. Associative desire is precisely the human desire to fend off alienation by creating rich relationships based on degrees of interdependence and mutuality; it represents the desire to know others and to be recognized as being integral part of a relationship, group, family, or community. Associative desire is the desire to be part of a collectivity greater than the self, a striving to be part of a larger identity. In addition, it represents the desire to express and receive empathy, to care for, and to be cared for, by others.

In contrast, liberal capitalist society, with its individualistic expression of desire, confines associative desire to the romantic private sphere, believing it 'unnatural' for people to truly desire association and cooperation within the public spheres of economics or politics. Whereas the Church attempts to mitigate this 'inherently' selfish nature through the obligation of charity, associative desire is generally regarded as inherently reserved for the private family or for those endowed with 'remarkable' altruistic abilities. A cooperative, associative desire within the social or political realms is regarded as the exception rather than the rule.

However, as anarchism and feminism demonstrate, we have the potential to express associative desire within both the public and private spheres by cultivating social relationships ranging from friendship and lovership to family, community, and political ties. Associative desire represents the potential which brings people to form culture and community, to participate in activities as diverse as joining clubs, attending parties, and engaging in politics. For better or for worse, most people have a desire to be in the presence of others, both in the intimate setting of friends and family and in the anonymity of the bustling city or market place. And in addition to constituting the basic desire for sociability, associative desire represents the creative striving toward greater levels of mutuality and cooperation: within the matrix of a cooperative community, people may create art, technologies, labor, relationships, and forms of self-government, centering such practices around the desire for mutualism and inter-dependence. Associative desire is the tendency to create social richness, to create non-hierarchical societies with mediated decision-making systems, complementary divisions of labor, and distributive economies.

In turn, associative desire moves individuals to cultivate structures which nurture the ability to express social desire. Associative desire is most easily expressed in contexts that are cooperative, non-hierarchical, and participatory. As social anarchism demonstrates, hierarchy and competition nurture social alienation, creating a climate of intimidation, mistrust, and animosity. In contrast, free from hierarchy and competition, people are better able to give each other the recognition, empathy, and attention that render life meaningful. Social anarchist and feminist structures which foster mutual aid and cooperation represent the associative dimension of the socio-erotic. Cooperative structures such as rotating leadership, collective ownership and labor, and direct participatory democracy represent but a few structural examples of the associative dimension of the socio-erotic within society.

Differentiative Desire: Knowing Self, Knowing The World

However, to fully actualize its liberatory potential, associative desire must be complemented by another form of desire, *differentiative desire* . Differentiative desire, the third finger of desire, is the desire to differentiate oneself within the context of a social group. Yet it also represents the desire to 'differentiate the world'—to make sense of the world through artistic or intellectual creative expression. Thus, while the first dimension of differentiative desire begins with the assertion *"I want to know myself,"* the second dimension begins with the assertion, *"I want to know the world."*

The first dimension of differentiative desire represents the desire to distinguish one's own identity within a wider social context. We may let the third finger of social desire be symbolized by the middle finger, representing the need to know and express the uniqueness of the self, to uncover one's particular efficacy, skill, strength, and potentiality. Differentiative desire rounds out associative desire by adding a complementary dimension of individuality. While we each yearn to feel part of a whole that is greater than ourselves, we also yearn to know and assert a self that is distinct within that greater collectivity. While associative desire represents a kind of 'urge to merge', differentiative desire represents a crucial 'urge to diverge' which allows an association to remain open to variation, innovation, and difference. Without the 'urge to diverge' of differentiative desire, an association is at risk of remaining static, homogeneous, and stifling.

The idea of differentiative desire could be termed the most 'Western' of the five dimensions of desire. In many cultures of the world people do not emphasize a notion of a 'self' that is separable from 'the people'. In fact, critics of Western societies often identify the idea of an 'individuated ego' as the cause of a lack of social humility and collectivity, qualities which are often associated with Asian, African, and indigenous cultures throughout the world. However, particularly within the liberal capitalist West, the idea of an undifferentiated self has often proven to be anything but liberatory. Paradoxically, although the idea of individualism is emphasized within the West, the idea of self-surrender is prominent as well. The fascist and nationalistic legacy of Europe illustrates the consequences of self-submission to a hyper-individuated authority or to the 'people', or *Volk*. As social anarchism demonstrates, Westerners must come to terms with the dangers of both hyper-individuation and hyper-association—expressions of selfhood that are equally capable of thriving within hierarchical and authoritarian societies. Both tendencies are capable of nurturing despotic abuses of and submission to authority.

Within the liberal capitalist West, association without differentiation enhances the likelihood of a mass of undifferentiated desires, increasing the possibility that individuals will join an association whose membership is predicated on expediency or the submission to religious and political charismatic authorities. In contrast, the 'urge to diverge' adds a complementary, liberatory dimension to associative desire which allows the self to be both collective and distinct. The desire to assert an innovative identity within a given collectivity allows for an open-endedness that is essential to the development of individuals and to the collectivity itself.

Feminist psychoanalytic theory has given significant attention to the potentially complementary relationship between associative and differentiative desire. According to Jessica Benjamin, each of us yearns to participate in what she calls "mutual recognition," a process in which two complete selves recognize each other as both dependent and independent. For Benjamin, the desire to both recognize otherness and to be recognized creates a dynamic tension which propels us to develop the capacity to recognize another person as a separate individual "who is like us, yet distinct."[10] For Benjamin, the idea of mutuality is predicated on this rich dialectic between two distinct selves rather than on a collapse of two selves into one.

Benjamin's notion of erotic "mutual recognition" differs dramatically from Freud's notion of erotic union. For Freud, union between individuals represents a desire "to make the one out of the more than one" in which the "more than one" represents a static totality, a suffocating unity that requires a negation of individual identity.[11] For Freud, because the self is inherently hostile to encounters with other distinct selves, erotic union requires the loss of self, permitting two identities to merge into one. Thus, for Freud, the desire to become one requires a unity achieved through the negation of self. In contrast, Benjamin's mutual recognition entails a unity *in* diversity. It implies a unity of distinct selves based on independence and interdependence. In turn, it implies a differentiation within association, a desire to maintain individual identity while recognizing a connection to others. Together, differentiative and associative desire can form an erotic dance between autonomy, community, individuality, and collectivity.

Differentiative desire is essential to true association *with* and to true differentiation *from* others. To know the particular ways in which we are distinctive, to understand our own complex motivations, dreams, and visions, allows us to 'get ourselves out of the way' when we seek to really see others. Paradoxically, knowing self allows us to really see and know others, for when we know ourselves, including our own prejudices, motivations, likes, and dislikes, we can see all that may obscure our ability to really recognize another person.

Whereas self-contemplation may represent a personal indulgence, authentic self-knowledge may serve a vital social purpose. For what we do not know about ourselves is potentially dangerous to others. For instance, in the case of racism or sexism, social ignorance can be lethal. What men do not know about the history of being men, or about their own socialization, or about how their desire for women has been constructed, may be dangerous to women. Most white people know little about the historical origins of their ideas of 'race' or 'whiteness,' remaining ignorant of the ways in which they benefit from and perpetuate hegemonic racist practices. Throughout history, the oppressed have always paid dearly for what the oppressors do not know about themselves.

In addition, what we do not know about ourselves is potentially dangerous to ourselves as well. Members of oppressed social groups are often deprived of knowledge of their own histories or cultures. This lack of self, or 'collective-self' knowledge destabilizes a group and makes it further vulnerable to social control. In contrast, self knowledge fortifies our ability to determine the degree to which we may be truly seen or known by another person. If we truly know ourselves, we are better able to assess the ability of another to perceive us accurately. In the same way, the degree to which we know ourselves heightens the degree of satisfaction we feel when another is truly able to see the qualities which render us utterly distinct.

Knowing The World

The second dimension of differentiative desire is the desire to know the world through creative and intellectual expression, to develop new ideas and art forms which give meaning to our lives, nuancing our understanding of the world. The ability to conceptualize is predicated on the capacity to translate abstract meaning into the differentiated forms of symbol or language. Differentiative desire is the desire to differentiate the world conceptually, making meaning where there was none before, to express our interpretation of reality. From the time we are children, we take great joy in finding the right words to describe a particular feeling. Language allows us to point to specific shades of meaning, allows us to experience the wondrous "ah-hah!" that emerges as we elaborate a theory that explains a mystery we might never have been able to articulate before.

Differentiative desire finds its expression in both the informal and formal philosophies of peoples all over the world. Although the mediums vary, the desire to differentiate the world through conceptual and verbal expression is a universal phenomenon. Language gives form to our ideas and feelings, allowing us to communicate the particularities of our experience. Through language, we may give shape to our experience and perceptions while also giving the world edges, texture, and meaning.

Historically, in the West, those in power have rigidly determined what would be defined as legitimate 'theory'. The most liberatory possibilities of the Enlightenment have too often been eclipsed by a capitalist tendency toward rationalization and instrumental logic. As many feminists, social ecologists, and indigenous theorists have demonstrated, the desire to differentiate the world solely through deductive, linear, or instrumental reason alone, has led to a way of thinking that is often reductive, fragmented, or relativistic. However, while breaking a subject down to its components can lead to a greater understanding of the whole, it can also fragment the whole into a sea of meaningless incoherent components. Hence, our desire to differentiate the world through ideas, language, and abstract conceptualization must also integrate an ethical associative moment: Through thinking associatively as well as differentiatively we give ethical coherence and unity to our thoughts as well.

While we may derive differentiative desire from the idea of association, differentiative desire also incorporates the idea of sensual desire. The 'sensual moment', we could say, is retained within differentiative desire. Although reason and sensuality are dualistically portrayed as 'opposites', theoretical engagement is often an intensely sensual event. As sensual, embodied beings, we may appreciate moments of pleasure that emerge as we articulate an elegant, well-crafted idea or argument. Sitting among friends, rapt in stimulating discussion, we may almost burst with the new idea percolating inside us. What could be more sensual than the great "ah hah!" that emerges from our throats when we finally grasp a new idea?

This 'sensual moment' surfaces within the act of artistic creativity itself. The artistic, creative impulse represents the desire to engender meaning and form that express something distinctive about the self or about the world. Differentiative desire represents the desire to use our senses aesthetically to express what is deepest within the human imagination, what tingles along the tips of our fingers. Few recognize the creative impulse to be as vital as the desire for sexual or sensual fulfillment; whereas it is expected that even the most 'average' person can achieve sensual fulfillment, it is rarely expected that each can achieve creative satisfaction through artistic expression. Creativity is reserved for the elite, regarded as a mere 'creative means' to an end that is generally quantified in terms of an economically valuable elitist 'product'.

However, the creative impulse need not constitute an instrumental means to an end. Creativity can represent a two-fold end in itself: the *expression* of a self, and another's *recognition* of this self-expression. In addition to yearning to creatively differentiate the world, we also long for the world to differentiate us, to distinguish us within the grand mosaic of life itself. In this way, the experience of both creating and being recognized brings fullness to creative self-expression. However, it is not necessary that our creativity be recognized as 'superior', awarding us social status, power, or profit. Rather, the acts of self-expression and recognition can be sufficient in themselves. While we long to be recognized as a part of an association, we also long to be recognized as distinctive within that association. In a free and cooperative society, creativity would become a dance of self-expression and recognition, reinforcing our sense of distinctiveness, community, and shared meaning.

Differentiative desire is the yearning to discover what is most distinctive about ourselves on an individual, community, or regional level. It is the desire to maintain and further elaborate personal and collective identity. And once we have identified what is most distinctive about ourselves, we often yearn to fulfill that distinctive potentiality. For instance, let us imagine being presented with the opportunity to learn to paint. Imagine that during this process we discover that we truly enjoy painting and that we find that we can paint particularly well. Indeed, we might yearn to further explore this particular form of self-expression. Differentiative desire represents the impulse to pursue all talents and abilities: social, creative, personal, and political. Differentive desire is the desire of the self to become more of itself: more complex, actualized, and elaborate than ever before.

Developmental Desire: The Desire To Become

It is here, at the conceptual boundaries of the differentiative moment, that the socio-erotic incorporates a developmental dimension. *Developmental desire*, the fourth finger, represents the desire to fulfill the distinctive talents or abilities which we uncover through the expression of differentiative desire. While we yearn to express who we *are*, we also seek to fulfill whom we *ought* to become as well. Developmental desire begins with the assertion "*I want to become.*" It represents the striving to bridge the gap between who we are at any given moment, who we could be, and who we *ought* to be—if we had the opportunity. Hence, developmental desire is symbolized by the pointer finger, the finger which points to the direction in which the self yearns to go.

In our society, developmental desire is often reduced to an instrumental motivation for the accumulation of power, status, or capital. Ironically, old people, who represent the elaborate and savory summation of a lifetime of differentiation and development, are largely regarded as "unproductive" unless they have accumulated a tremendous amount of capital over the years.[12] However, despite this narrow view of human development, the desire to develop endures. Developmental desire resurfaces as the relentless craving of the individual to uncover distinctive potentialities and as the collective desire of society to unfold its distinctive possibilities as well. The desire to develop emerges as a restless apprehension; a desire to taste possibility on the tip of our tongues, unable to rest until we taste more.

In addition to differentiating ourselves to uncover the widest spectrum of creativity, sensuality, empathy, and personality, we also yearn to grow *developmentally.* In this way, development is linked, but not reducible, to differentiation. Understandably, many confuse change, growth, and variation with development. We reason that by differentiating ourselves from a particular time, place, or identity we will develop, mature, or 'evolve'. However, rather than cultivate degrees of maturity or coherence, we may achieve a differentiated *stasis.* We may have changed our show and taken it on the road, only to find that the road is winding in circles. Hence, differentiation is not equivalent to development. In the case of multiple personality disorder, an individual unconsciously responds to trauma by splitting the personality, differentiating the self into a myriad of sub-selves, each of which endures and copes with the stress and pain of abuse. In this instance, while the self succeeds in the task of differentiation, it fails to develop into a coherent unity. As a result, an individual suffering from this disorder serves as a host to a diversity of differentiated sub-selves, each lacking the unity and maturity necessary for true development and integration.

Developmental desire is precisely the desire of the self to become increasingly unified within the diversity of its own differentiation. For instance, while we may wish to uncover our distinctive potentialities for creativity, sensuality, and cooperation, we also yearn to discover an overriding logic that can endow our lives with meaning and wholeness. We can all think of someone in our lives who possesses a myriad of interests yet is incapable of focusing long enough to sufficiently develop a single one. We would say that their focus lacks the very unity or coherence necessary for self-development. In this way, whereas differentiation rounds out the idea of association, development rounds out the idea of differentiation, adding to it a dimension of unity necessary to make the self not only diverse, but dynamic, whole, and meaningful. Hence, development is qualitatively different than a mere process of change or growth. According to Bookchin, the often painful dialectic of a developmental desire is necessary for the differentiation or maturation of the self:

> Desire itself is the sensuous apprehension of possibility, a complete psychic synthesis achieved by a "yearning for..." Without the pain of this dialectic, without the struggle that yields the achievement of the possible, growth and Desire are divested of all differentiation and content.[13]

So far, we have been exploring the idea of development on an individual level. Yet such a utopian understanding of development may be applied to society as well. Each society has the potential to express its collective developmental desire to become increasingly differentiated and whole. However, under capitalism, the naturalistic metaphor of 'growth' is deployed to naturalize the immoral hoarding of capital. Within the social Darwinian view of development, the 'fittest' that survive are those who accrue the most profit and power. Few expect society to become ever more differentiated, dynamic, and whole. Rather than being evaluated qualitatively, social development is measured quantitatively as the growth of capital itself. Developmental desire is reduced to the individual desire to differentiate one's self from the masses through the accumulation of capital and social status.

This individual desire is then 'collectivized' into the shared desire of most Americans to distinguish themselves from those of 'less developed' Third World countries. Meanwhile, this social arrogance is predicated on a capitalistic idea of 'growth', obscuring a true understanding of development as an incremental process in which individuals and society may become qualitatively richer, developing deeply textured capacities for empathy, interdependence, and creativity.

Hence, the idea of 'growth', individual or social, is insufficient for cultivating a full understanding of development. As we have seen, true organic development is a process of differentiation and wholeness. In turn, this development entails the act of becoming which is distinguishable from the simple idea of growth. For instance, when a seed unfolds into a flower, the seed does not merely 'grow' or become a bigger seed. If development were simply growth or expansion, then there would be no flowers at all, just gargantuan seeds swaying in the fields. Instead, something dramatic occurs within the logic of the seed; something within the seed's very structure allows it to differentiate into a new, more elaborate form. The seed gradually gives way to the flower not merely by expanding but by differentiating into an ever more complex organism. This dialectical process of becoming moves from the first thread-like root of the seedling to the upward rising of the stem through the gradual maturation and emergence of the blossom itself. Through this development, the seed is not destroyed; rather, it unfolds within the logical progression of its own internal structure. In this way, we could say that there was something distinctive about the seed's structure which allowed it to engage in this process of 'becoming', undergoing a series of phases in which it was able to become 'more of itself'. We could say that the flower represents the differentiated expression of the seed's potential for becoming a flower.[14]

In contrast to this social ecological view of development, capitalist society regards development as hierarchical, competitive and determined. Under the rubric of liberal capitalism, to differentiate means to separate and surpass what we were before, assuming a state of superiority over others. Such an approach to development emerges within the deterministic models of development proposed by thinkers such as Hegel or Marx. Whereas these thinkers contributed immeasurably to the world of dialectics, offering an understanding of the logical unfolding of symbolic and material reality respectively, their dialectical approaches retained a determinism that must be transcended. Both thinkers portrayed development as a series of *necessary* negations: a linear and hierarchical process in which earlier phases of development are necessarily overcome by 'superior' later phases. According to Hegel, whereas change is made possible by the process of contradiction and negation, conflict and opposition represent the only means by which development may occur; thus, out of the bland, static world of 'being' emerges the oppositional, dynamic world of 'becoming'. In order for a thing to become something else, it must overcome that which preceded it.

Similarly, Marx regarded the development of society as a series of necessary negations. For Marx, whereas earlier 'primitive' societies must be overcome by increasingly rational and civilized societies, social history represents an inevitable linear trajectory. Beginning with so-called primitive societies that become increasingly technological, hierarchical, and competitive, history finally gives way to a free and socialist society. In this way, Marx ascribed to a liberal notion of 'progress', asserting the necessity of hierarchical systems such as capitalism as a stepping stone toward a higher expression of civilization. Moreover, in the same way, Freud follows in this tradition, regarding child development as a series of self-negations or repressions. Whereas 'maturity' is marked by a negation of earlier impulses and desires, Freud's 'rational adult' marks the pinnacle of white male self-repression.

However, the 'history of society', is not a singular or monolithic event. Society and culture develop in different locations, fashions, and times. Each society must be understood integrally as the summation of its own historical development. Furthermore, the process of social development is uneven; within a given society, there may be particular cultural or political practices that are more complex and developed than others. For instance, while one culture may develop a particularly sophisticated system of agricultural or industrial technology, that same culture might be marked by a particularly 'maldeveloped' form of governance incorporating violence, dominance, and rigid social stratification.[13] Similarly, while one society may practice particularly laborious systems of agriculture, that same society may have developed intricate systems of self-government, nuanced in their degree of non-hierarchy, complementarity, and cooperation.

In contrast, new 'organic' dialectical thinkers such as social ecologist Murray Bookchin and psychoanalyst Jessica Benjamin propose an alternative view of development. Indebted to Hegel, both thinkers regard development as cumulative, depicting later phases of development as incorporating earlier ones and bringing them to a level of more complex differentiation. However, for Bookchin and Benjamin, this crucial 'negative moment', inherent within all processes of development, is mediated by the idea that development may be cumulative, cooperative, potential, and open-ended rather than determined and hierarchical. Bookchin and Benjamin elaborate upon what is best within Hegelian 'negativism' by drawing out a more organic and non-hierarchical view of development.

For Hegel, when a self recognizes itself as separate from another self, it will strive to annihilate the other. For Hegel, social relationships are inherently marked by a conflictual struggle for power in which individuals vie for attention and recognition, generally ending in a one-up situation. In contrast, Benjamin asserts that the self may potentially yearn for the presence of others out of a desire to develop. For Benjamin, development does not occur despite others, but because of others: the relationship between an infant and mother is potentially mutually beneficial rather than inherently conflictual.[16]

According to Benjamin, development occurs within a social context, preferably within a context that nurtures both individuality and connection. Rather than constitute a series of negations, development represents a series of increasingly complex expressions of relatedness and individuality. For instance, a child does not necessarily have to separate from its mother in order to mature. Rather, it may differentiate itself within that relationship, developing an increasingly nuanced ability to be both related and independent, both recognizing and being recognized by its mother. In this way, Benjamin introduces the idea that development may be a cooperative, dialectical process in which latent abilities for independence and dependence are developed and expressed.

In addition to being marked by accumulation and cooperation, human development can be marked by open-endedness and non-determination. For instance, at birth, each individual represents a series of biological and environmental 'givens'. In turn, there exists a degree of chance, or spontaneity, that informs how these 'givens' will be organized and how they will evolve. Biological and environmental factors, then, represent a set of *potentialities* rather than a set of determinants. There exists no determined blueprint which guarantees how an individual will *necessarily* develop, or whether they will develop at all. Organic life is marked by a dimension of potentiality which provides a horizon of logical yet undetermined possibilities that may or may not unfold.

Developmental desire is precisely the desire to develop the particular spectrum of 'logical possibilities' that exists within each of us. It is the desire to participate actively in our own development, differentiating ourselves into what we could be, bringing ourselves to a new level of complexity and integration. Developmental desire is *not* the desire to develop our abilities to dominate or master our earlier or less mature impulses; rather, it is the desire to integrate our earlier 'child self' with our emerging 'adult self'. When this integration is achieved, we are able to retain levels of spontaneity, flexibility, and authenticity characteristic of the child, integrating these qualities into the cognitive, self-reflexive, and empathetic capacities of adulthood.

We long to differentiate ourselves, to coherently unfold what is distinctive within us. We yearn as well to develop cooperatively in a spirit of open-endedness and possibility rather than in a spirit of reductive determination. Instead of merely striving to accumulate capital or power, developmental desire represents the desire to develop qualitatively, to lead richer, more meaningful lives. Within a free society, developmental desire represents the motivation that propels individuals and society toward an open horizon of unending development.

However, within the context of liberal capitalism, the full range of cooperative and creative potentiality lies largely undeveloped while a narrow spectrum of competitive and instrumental abilities are nurtured to extremes. Even within this narrow range of 'acceptable potentialities', it is mainly the most privileged who gain access to the material means by which to develop their abilities, be they intellectual, athletic, artistic, or even the more instrumental abilities such as state politics or business. Hence, we might ask ourselves: what happens to developmental desire in a world which eclipses its utopian potential?

Oppositional Desire: The Desire To Fight Injustice

To explore the fate of developmental desire within the context of social hierarchy, we must uncover within the socio-erotic an oppositional dimension that may potentially emerge as we confront obstacles that impede our full individual and social development. Oppositional desire, the fifth finger of social desire, represents the rational inclination to oppose all individuals, institutions, and ideologies that obstruct the full expression of all forms of social desire, be they sensual, associative, differentiative, or developmental. Oppositional desire may be symbolized by the open palm. This first moment of opposition, the moment of critique, represents the act of rationally reflecting upon that which obstructs our expression of other forms of social desire, analyzing the history of oppression, and reasoning out coherent plans for future resistance. When we (metaphorically) 'read' the receptive and integrative palm, we know when to oppose even the desire for opposition, recognizing the appropriate time to wait, listen, and be critical, holding the serious and specific weight of the world in our open hand.

However, opposition cannot be waged by contemplative critique alone. When the five fingers of desire come together, they also form a fist of collective or individual defiance. This second moment of opposition, then, the moment of resistance, represents taking passionate and rational action to defy institutions that impede the creation of a just new world. Such acts may be covert or overt, or they may assume the form of armed insurrection or active non-violence. Throughout history, wherever there is a story of oppression, there is a hidden and unspoken story of resistance. The oppositional desire of the fist held high symbolizes the unity and strength of social and political contestation.

Finally, opposition requires a third, reconstructive moment. Oppositional desire may be symbolized by the 'opposable thumb' that brings reconstructive and evolutionary possibilities into being through critical invention. Often, the desire for resistance is the mother of invention as oppressive circumstances inspire us to imagine and reason new ways not only to survive but to flourish. Opposition is incomplete without the act of reconstructing a coherent and organically rational vision of the future. It is insufficient to merely critique and contest social and ecological injustice. Opposition enters into its fullness when we begin to think through our oppression to create a desirable new world.

The expression of oppositional desire can be suppressed by authority, but it cannot be dissipated altogether. Moments of overt oppositional desire emerge in the direct demands for freedom that make up the body of social demonstrations and resistance throughout history. However, oppositional desire cannot always be expressed overtly. Sometimes, it will assume covert forms ranging from anonymous acts of sabotage to the most subtle expressions of psychological resistance. The socio-erotic, then, represents not just the overt expression of a range of social desire. It also represents the *potential* for social desire, the impulse toward freedom itself. Oppositional desire is the force that pushes green tongues of weeds through cracks of the blandest parking lots, just to say: "I will not go away." It is that which inspires us to resist, not just to fulfill our basic material needs, but to express our desire for a particular quality of life, a particular sensuality, connectedness, and texture that endows life with meaning and a deep sense of satisfaction.

Five Qualities Of Oppositional Desire

All five fingers of social desire can be rendered oppositional in a context of social hierarchy and oppression. For instance, sensual desire may assume an oppositional dimension when we oppose forces which obstruct our desire for sexual or sensual self-expression. Women's fight for sexual freedom represents a form of oppositional sensual desire as women fight for the right to love and determine the fate of their own bodies. The movement for lesbian, gay, bisexual, and transgendered liberation represents moments of overt oppositional desire when people take action to challenge patriarchal institutions of compulsory heterosexuality. Sensual desire assumes an oppositional dimension when we incorporate our love for beauty into forms of direct action, creating new ways to express dissent and visions of a utopian future through visual art, theater, music, and poetry. The desire for 'nature', when expressed in oppositional terms, represents as well an expression of oppositional developmental desire. The yearning to restore and elaborate ecological integrity by contesting capitalist and state practices, and the desire to fight the parallel social and ecological injustices that constitute environmental racism reflect what happens when the social desire for 'nature' encounters moments of ecological injustice.

The second finger of social desire, associative desire, may assume an oppositional dimension when we resist forces that obstruct cooperation. Resistance to oppressive institutions such as racism, sexism, and capitalism, which counter the desire for mutual recognition, is born out of associative oppositional desire. In turn, social experiments in intentional communities, or worker-collectives, represent examples of associative oppositional desire. Attempts to share, barter, or cooperate when such activities are discouraged or prohibited, demonstrate the relentless socio-erotic opposition to the institution of capitalism. In addition, when people risked their lives to work the underground railroads or to hide slaves in the U.S.; when a battered woman runs to a phone booth in the middle of the night to call a friend; when a poor woman gives her neighbor money for food, such acts represent expressions of the desire to oppose through association, pushing past institutionalized sources of separation, isolation, and alienation.

The third finger of social desire, differentiative desire, may become oppositional when we are confronted by systems of authority that demand expedience and conformity. Oppositional differentiative desire is the push to differentiate our own desire from the desire of those in power. Within the context of hierarchy, differentiative desire takes on a new impulse. Rather than differentiation *within* the context of a greater cooperative collectivity, differentiative desire becomes the desire to differentiate *from* the ideas, institutions, or individuals in power.

Sabotage, often misinterpreted as self-defeating behavior, can represent a vital act of self assertion. Just as men may misinterpret women's sexual desire as 'irrational', they may misinterpret women's oppositional desire too, misperceiving women's resistance as 'incompetence'. In *Lesbian Ethics*, Sarah Lucia Hoagland discusses Donna Deitch's documentary *Woman to Woman*, in which a working class housewife describes feelings of frustration and helplessness in regard to her life and work within the home.[17] At one point in the interview, the woman gets a gleam in her eye, lowers her voice, and asks the interviewer, "Have you ever bought something you don't need?" Confessing to the interviewer that she often buys cans of beans she has no intention of using, just to waste her husband's money, she concludes, "You have to know you're alive; you have to make sure you exist."[18]

This desire for agency or self-determination is an act of oppositional differentiative desire. This desire is expressed in a spectrum of sabotage activities ranging from burning dinners to hiding the master's tools on the plantation. As Hoagland points out:

Acts of sabotage can function to establish that self, to affirm a woman's separateness in her own mind. It may be more important to the woman who burns dinners to remind herself (and maybe her husband) that he cannot take her for granted than it is for her to rise socially and economically...And it may be more important to the slave that she affirm her existence by thwarting the master's plan in some way than it is to secure safety in a situation in which believing she is safe is dangerously foolish. If a woman establishes her self as separate (at least in her own awareness) from the will of him who dominates by making certain decisions and carrying them out, then those choices are not self-defeating, since without them there would be no self to defeat.194

Differentiative desire lies at the heart of oppositional desire. Through opposing the power which oppresses us, we differentiate ourselves from that power, asserting our independent desire for freedom. Often, the cost of differentiative oppositional desire is our own physical defeat, a sacrifice that challenges an exclusively materialist interpretation of social resistance. Predicating social and political resistance on material necessity alone can never account for the ways in which the subjugated often forgo their own physical security, safety, and even survival, in order to maintain an integral sense of selfhood and community.

The fourth finger of desire, developmental desire, assumes an oppositional dimension when confronted by obstacles to self-development on an individual, social, or community level. Social hierarchy functions to stay the development of those at the bottom. This 'pressing down' on individual and social development takes place on levels that are physical, emotional, intellectual, spiritual, ethical, and creative. For instance, within many capitalist cultures, women are de-skilled technologically and intellectually, instilling a lack of confidence and competence in abilities that should belong to both sexes. Within liberal capitalist societies, knowledge regarding such areas as sexuality, health, and technology is often stolen from women and other oppressed peoples to be hoarded and controlled within centralized institutions such as hospitals, universities, corporations, and governments.

In the crisis today over intellectual property rights, First World corporations steal and patent seeds cultivated over thousands of years by indigenous peoples in the Third World. The goal of such capitalistic exploits is to centralize the cultivation and distribution of seeds, de-skilling local farmers in the process. Unless interrupted, such action threatens to erase not only local agricultural knowledge but the communal and historical development of agricultural knowledge itself.[20]

In addition, while the oppressed are often de-skilled, they are also taught to forgo their own developmental desire. In many societies, women are encouraged to engage in vicarious expressions of desire, nurturing the development of children and men. Viewed as less developed than the mature male capitalist subject, for example women are often described as being closer to nature; a nature that is in turn portrayed as lowly and static, deprived of developmental, self-organizing properties. Accordingly, women, like nature, must await the 'animating principle' of man and his technology and intellect in order to develop or grow. As Simone de Beauvoir points out in *The Second Sex*, only elite modern man can ever hope to gain development, or transcendence over the alleged stasis and repetition of the natural world.[21] Women and the rest of the oppressed must remain within immanence, or within a state of unending latency, without any hope for development.

Developmental desire becomes oppositional when people begin to acknowledge and elaborate the development which they have achieved. In 1936, the "Mujeres Libres," an anarchist organization of "free women" who fought in the Spanish Civil War, established self-development as a central focus in women's revolutionary work. Like most social anarchists, the Mujeres Libres regarded the transformation of the self as crucial to the transformation of society.[22] Transcending a Marxian oriented 'needs agenda,' the Mujeres Libres asserted women's desire for social freedom, working to develop new skills and abilities while fighting to create a qualitatively new society. In particular, the Mujeres Libres established *capacitacion*, an agenda which prepared women for revolutionary engagement, and *captacion*, which incorporated women into the libertarian movement. This dual orientation was expressed clearly in its statement of purpose:

> ...to create a conscientious and responsible female force [originally, a "revolutionary force"] that can act as a vanguard of progress; and to this end, to establish schools, institutes, conferences, special courses, etc., designed to empower women and emancipate them from the triple enslavement to which they have been, and continue to be, subject, the enslavement of ignorance, enslavement as a woman, and enslavement as a worker.[23]

Through an agenda of *captacion*, women focused on developing their participation within anarchist organizations. Due to the widespread neglect of women's issues by the larger anarcho-syndicalist movement, the Mujeres Libres addressed social and economic oppression that specifically affected women, working to overcome those obstacles, to integrate women into the wider revolutionary movement.[24] In turn, through *capacitacion*, women expressed their desire to reestablish their capacities for both social and self renewal.

While their education focused primarily on the areas of literacy and sexual education, they emphasized as well a wide range of other skills that would prepare women for their life and work in the new anarchist society. In the Fall of 1936, the Mujeres Libres in Barcelona offered intensive courses in general culture, social history, economics, and law in its offices in the Plaza de Cataluna. Regardless of the topic, the theme was the same: Women must take responsibility for their own development, education, and participation within the larger movement.

Oppositional developmental desire has continued to surface throughout history as people challenge conditions of personal and collective stasis caused by oppression. During the New Left for example, feminists established a 'developmental' agenda, creating consciousness-raising groups designed to allow women to increase awareness of oppressive gender roles. In turn, in the Third World, beginning in 1977, women have expressed developmental oppositional desire in the "Green Belt Movement" in Kenya. In this movement, activist and scholar Wangari Maathai formed a network of grassroots educational and activist groups throughout that country to prepare women to address the parallel crises of deforestation and poverty. Training women to work in such areas as seed cultivation, marketing, and forest management, The Green Belt Movement restored green areas around school compounds and city limits throughout the country. Seeking more than ecological and economic restoration, however, The Green Belt Movement allowed women to develop their status as holders of expert knowledge.[25]

Expressing oppositional desire is a way to feel alive in a world which deadens our yearning for freedom. To resist, on an individual and social level, is vital to the revolutionary project; when people forget that they possess the very means for social change, they become ignorant of their own potential for dynamism and self-development. When people see themselves as 'stuck', they are likely to believe that the world is inevitably unchangeable as well. When we lose confidence in our ability to develop new oppositional ways of being, we lose faith in our ability to change the world. Propelled by our oppositional desire, we have the potential to challenge the 'big lie of stasis' that teaches us that the world is controlled by an unchanging set of natural laws that keeps each thing and each person in place. Once we recognize that we can fight oppression to become more sensual, cooperative, creative, and whole, the big static book of natural law looses its yellowed pages as they scatter in the winds of opposition.

The Socio-Erotic: Toward An Informed Desire

The five dimensions of desire provide a way to talk about qualitative dimensions of reality without appealing to spiritual or purely intuitive explanations, a way to translate that which is conventionally called spiritual into that which is erotic. Yet such an approach requires a re-thinking of vernacular understandings of meaning that are commonly contrasted against the idea of reason. In a world of capitalist rationalization, a world that reduces social and ecological relationships to standardized units of profit, it is tempting to appeal to ideas of sacredness or spirit to convey the poetry of life, dimensions of reality that cannot be reduced to instrumental or linear reason. However, when we equate all that is rich, deep, and intensely meaningful with that which is not rational, we conflate rationalization with rationality, failing in turn to recognize moments of organic rationality and history within what is usually invoked as spiritual. We fail to realize that we can use reason to create structures and ways of being that are intensely meaningful in the most cooperative and liberatory sense.

Again, the desire to assert a dimension of life that cannot be bought, sold, or biologically determined moves us to embrace the idiom of spiritus rather than that of rationality or cooperative relationality. Believing that rationality is inherently reductive, we posit the poetry, sensuality, and inter-relatedness of life as a kind of universal essence or energy that flows through the world, a kind of activating principle that is beyond history or reason. Yet there are other more relational, rational, and historical ways to describe moments of holism, ways to articulate instances in which the whole cannot be reduced to a mere sum of its parts. The idiom of the socio-erotic provides a way to point to such qualitatively irreducible moments, re-configuring the dimensions of social desire as social rather than spiritual or intuitive, rational rather than irrational, historical rather than universal, and common rather than sacred.

The socio-erotic, then, provides a way to talk about that which is rational and irreducible, that which is poetic *and* rational, historical, and social. Social desires are marked by moments of rationality or logic that are reflective of the historical, social, and political contexts in which they emerge.[26] In this way, the socio-erotic is not a universal, irrational essence or spirit; rather, it represents a way to talk about a range of social desires that are informed by and answerable to historically situated cultural practice. Moving beyond essentialist ideas of spirits, energies, forces, or drives, we may uncover the most meaningful and social implications of cooperative relationality itself.

As social creatures, than, our most meaningful and cooperative social yearnings are marked by an underlying rational, historical, and relational logic. When participants in the civil rights movement yearned for social justice, for example, such yearnings were not a priori, or instinctual. Instead, they reflected historically rooted and rational understandings of what ideas of 'race', 'justice', and 'injustice' meant during the post-war period of post-slavery America. The social desire articulated through the poetic prose of James Baldwin reflects a highly rational mind capable of articulating compelling arguments against racism and heterosexism in a language of sensuality and profound emotion. Baldwin's creativity cannot be explained as a simple energy, force, or drive, but as an expression of a particular relationality: a meditation upon a rich matrix of social and political relationships that Baldwin observed, lived, and reflected upon in a particular place and time in history. By describing the social desire of Baldwin as a merely intuitive expression, we miss the profoundly historical, rational, and relational nature of this artist's work.

Articulated through the language of the socio-erotic, we may see moments of sensual desire in Baldwin's prose: a relational desire for a quality of mutual recognition that countered racism, classism, and heterosexism. Baldwin expressed a rational desire for association in his discussions of brotherhood, unity, love, and compassion. Yet again, rather than represent essential intuitions or an expression of spirit, we may recognize within the genius of Baldwin the ability to seamlessly join a critique of political and social structures with a plea for a sensuous expression of human compassion and unity. Baldwin's reflection upon his own thirst for creativity, sensuality, and knowledge as a young black man in Harlem in the 1940s, a desire that sent him to the public library, to the pulpit, and into the arms of young men, represents not an irrational spiritual drive or intuition but a highly rational and historically situated expression of a relational differentiative and developmental desire. In turn, Baldwin's writings against racism represent sensually articulated expressions of oppositional desire, a desire that is impassioned and marked by an undeniable logic:

> At bottom, to be colored means that one has been caught in some utterly unbelievable cosmic joke, a joke so hideous and in such bad taste that it defeats all categories and definitions. One's only hope of supporting, to say nothing of surviving, this joke is to flaunt in the teeth of it one's own particular and invincible style. It is at this turning, this level, that the word color, ravaged by experience and heavy with the weight of peculiar spoils, returns to its first meaning, which is not *negro*, the Spanish word for black, but vivid, many hued...the rainbow, and warm and quick and vital...*life*.[27]

To attribute Baldwin's genius to spiritus denies the distinctly embodied, historical and *human* quality of this work. By identifying Baldwin's genius as an expression of social desire, we may reclaim an appreciation of the human potential for making liberatory, creative, and meaningful connections out of the matrix of social relationships themselves. We may indeed describe Baldwin's work as *socio-erotic*.

Yet recognizing the historicity and sociality of our social desires does not imply that we should rationalize or reduce such experiences to behaviors that are operative, biologically determined, or merely socially constructed to fulfill some adaptive function. Appreciating the socio-erotic does not entail that we become self-conscious each time we engage in meaningful activity, wringing the poetry out of each experience by analyzing its rational and political implications. To be sure, there are some experiences that are degraded by in-the-moment analysis: The poetry of sexuality, artistic expression, and parental love, for instance, may be compromised by constant appeals to critical self-reflection. What makes a particular song beautiful or pleasurable is often the ability to temporarily lose or suspend self-awareness, allowing the self to dissolve into a delicious rhythm. However, it is naive and perhaps even dangerous to think that because we can suspend awareness of the rationality or history underpinning such experiences, because we can shift awareness away from what it is that makes us label a particular song, face, or mountain as beautiful, that those inscriptions of what is beautiful stand outside the realms of rationality or history.

Assertions of irrationality or intuition as epistemologically more authentic or immediate than reason are predicated on the myth that reason is the opposite of intuition. However, intuition often constitutes a pre-reflexive expression of rationality: when intuitions are right, they reflect historically grounded insights that we have rationally cultivated about the world; when they are wrong, they often reflect more about ourselves and our unconscious desires. Intuitions can, indeed, often be wrong and destructive: Whereas Hitler intuited that the Jews were a sub-human enemy to the German *Heimat* or homeland, and anti-abortionists intuit that first trimester fetuses are 'babies' that should be protected, there exist many men who intuit that their wives are unfaithful, and deserve a beating. Conversely, many intuitions, defined as irrational, or pre-rational, are often grounded in highly refined bodies of local knowledge. So often throughout the history of the patriarchal and colonial West, 'women's intuitions' and indigenous 'folk knowledge' are cast as irrational to dismiss highly rational understandings of human behavior and natural processes.

The Enlightenment's failure to transcend misguided and solipsistic views of rationality, views that often dismissed the rational knowledge of the marginalized, may inspire us to cultivate new ways of approaching questions of rationality so central to feminist and subaltern epistemology. As we reject reductive discussions of rationality, we may engender epistemological options beyond appeals to spirituality and intuition. The idea of the socio-erotic represents an embodied and historical approach to questions of meaning, connectedness, sensuality, development, and moral opposition. A rationally informed social desire, a desire informed, provides a radical new approach to such crucial questions, so central to the social and ecological struggle. The socio-erotic provides a metaphor that better resonates with the shift from a spirituality-based essentialism to a historically situated relationality.

By appreciating the meaning of the socio-erotic, the dimensions of social desire, we valorize the immense beauty, power, and intelligence that marks our most sensual, empathetic, and developmental ways of relating. Far from being reductive, we may elaborate an appreciation for the stunning potential of humanity to express its relationality in sensual, creative, and dynamic ways. Thus, if the socio-erotic is the oppisite of anything, it is not spirituality or the sacred, but to capitalist rationalization, and an anti-humanism that reduces humanity to a cold and controlling anti-social species: a portrayal that dismisses and trivializes the potential of humanity for engendering institutions that nurture the most empathetic and sensual expression of social and ecological relationships.

By viewing meaningful experiences through the lens of the socio-erotic we regain a poetic appreciation of the diverse expressions of human sociality. We root our goodness not in spirituality or in romantic purity, but in our humanness, a humanness that is derived from and constituted by, natural history itself. It is deeply radical to assert what is potentially good in humanity during cruel and truly anti-human times such as these. In a neo-liberal era in which the majority of humanity is exploited, despised, and tyrannized, it is an act of the greatest empathy to recognize within those who are not free, the potential for beauty, intelligence, cooperation, and freedom.

In an era dominated by Christianity and neo-liberal capitalism, it is tempting to yield to portrayals of a humanity that is inevitably flawed, selfish, and ecologically destructive, a species inherently opposed to an innocent and pristine natural world. The anti-humanism that pervades the radical ecology movement, an anti-humanism that encodes 'knowledge' and 'rationality' as sinful or regressive, perpetuates the religious myth of a world that 'fell' because of humanity's quest for knowledge and pleasure. In turn, the romantic idealism that marks ecological discussions encourages us to idealize 'nature' (while hating our 'flawed' selves) rather than resist social institutions that allow the anti-social few to degrade the rest of humanity and the natural world. Ecological romanticism allows us to keep social hierarchies intact, constructing idealized 'nature preserves' or 'natural products' for the pleasure and guilt reduction of the privileged few.

The socio-erotic represents the attempt to further differentiate the idea of social desire, differentiating in turn, the cooperative impulse itself: elaborating the desire for mutualism and an ethical and oppositional progression toward a utopian horizon. Our vocabulary for describing moments of desire has been impoverished for centuries; indeed, it has been limited to the language of energistic, individualistic, and romantic drives for material acquisition, status, and personal sensual pleasure. We need to develop a new language of desire, offering ourselves a broader palette of colors to paint ever finer shades of meaning, subtlety, and nuance. Thinking through the socio-erotic represents one step toward developing this language, moving us toward a greater fluency in the language of freedom itself. We need to rationally fall in love with what is potentially most empathetic and progressive within social relationships. By focusing on the quality of *relationships* to self, others, and to the rest of the natural world, we move away from appeals to universalizing *essences*, to articulate crucial cultural meanings and social relationships. Trusting ourselves to think compassionately, organically, and relationally, we may take the apple of knowledge into both hands and bite down hard.

Notes

1. Benjamin Barber, in his book *Jihad. vs. McWorld,* elaborates upon the idea of McDonald's as a metaphor for the mood and mechanism of 'advanced' capital. For Barber, the parallel emergences of global capital and religious fundamentalism represent a paradoxically complementary threat to democracy itself. See Barber, *Jihad v. McWorld* (New York: Random House, 1995). Also, for a truly stimulating discussion of the meaning of service economy in an era of 'flexible accumulation, see David Harvey's *The Condition of Postmodernity* (Cambridge: Blackwell, 1990).

2. Max Weber initiated a century-long discussion of the idea of 'disenchantment'. The term 're-enchantment' was popularized by students of Weber, members of the Frankfurt School including Max Horkheimer, Theodore Adorno, and Herbert Marcuse. Both terms have subsequently captured the imaginations of a range of theorists engaged in postmodern and ecological discourse, thinkers searching for a way to talk about the erosion of meaning and ecological integrity within modern and postmodern capitalism.

3. See Janet Biehl and Peter Staudenmaier, *Ecofascism: Lessons from the German Experience* (London: AK Press, 1995).

4. The question of 'mystery' has dominated much discussion in feminist and ecological circles. Rightly dismayed by reductive analytical reasoning that reduces phenomena to meaningless fragments in the pursuit of 'rational knowledge', many thinkers have advocated embracing the idea of 'mystery' as a way to point to moments of irreducible meaning. Such discussions have led to pleas to put 'mystery' back into politics as a way to 're-enchant' an otherwise instrumental political practice. For a brief discussion of 'mystery', see Ynestra King, "The Necessity of History & Mystery," in *Woman of Power* 1988.

5. The modern transmogrification of *Eros* (a pre-Olympian deity who, born of Chaos, personnified love in all of its aspects) into an 'energy' or 'Life Force' is most closely associated with Sigmund Freud. Modeling the human psyche after the steam pump, Freud described the psychic world as a mechanism analogous to a series of pressure chambers activated by the fluctuating pressure and release of steam energy. Freud transformed the mythological narrative of *Eros* into this mechanistic model, establishing *Eros* as a 'steam-like' impulse, energy, force, or drive that would propel social behavior. Also, for a more historical and social discussion of Eros, see Herbert Marcuse, *Eros and Civilization* (Boston: The Beacon Press, 1955). Although Marcuse retains the energistic approach to Eros taken by Freud, he pioneered a discussion of Eros as a potentially constructive social impulse.

6. According to Nicholas Xenos, it is within classical liberal theory that we first see an explicit theory of scarcity associated with ideas of need, desire, individualism, and capitalism. See Nicholas Xenos, *Scarcity and Modernity* (New York: Routledge, 1989).

7. For a discussion of the emergence of sexual discourses in Western history, see Michel Foucault, *History of Sexuality, Vol. 1* (New York: Vintage Books, 1980).

8. For a critical examination of technology within contemporary ecological discourse, see Murray Bookchin, *Re-Enchanting Humanity* (London: Cassell, 1995).

9. See Donna Haraway, *Simians, Cyborgs, and Women: The Reinvention of Nature* (New York: Routledge, 1991).

10. Jessica Benjamin, *The Bonds of Love: Psychoanalysis, Feminism and the Problem of Domination* (New York: Pantheon Books, 1988), p. 23.

11. See Sigmund Freud, *Civilization and Its Discontents* (New York: W. W. Norton & Company, 1961).

12. Lizzie Donahue. Personal Communication. 26 April 1995.

13. Murray Bookchin, *Post-Scarcity Anarchism* (Montreal: Black Rose Books; reprinted 1986), p. 302.

14. In the *Philosophy of Social Ecology*, Bookchin provides an in-depth examination of notions of organic development from a dialectical perspective. See Bookchin, "Thinking Ecologically," in *The Philosophy of Social Ecology: Essays on Dialectical Naturalism* (Montreal: Black Rose Books, 1995).

15. As social ecology and ecofeminism demonstrate, the idea of "modern development" is markedly biased by a capitalistic interpretation of society and nature. As Vandana Shiva illustrates, the capitalist interpretation of development represents a "maldevelopment" based on unrestrained economic growth, predicated on the work of women and the Third World itself. See Vandana Shiva, *Staying Alive: Women, Ecology and Development* (London: Zed Books, 1989), pp. 5-6.

16. Jessica Benjamin, *The Bonds of Love*.

17. Sarah Lucia Hoagland, *Lesbian Ethics: Toward New Value* (Palo Alto: Institute of Lesbian Studies, 1988).

18. Ibid., p. 40.

19. Ibid., p. 47.

20. See Vandana Shiva, "The Seed and the Earth: Biotechnology and the Colonisation of Regeneration," in *Close to Home*, ed. Vandana Shiva (Philadelphia: New Society Publishers, 1994).

21. Simone de Beauvoir, *The Second Sex*, 2nd ed. (New York: Vintage Books, 1952).

22. Martha Ackelsburg, *Free Women of Spain: Anarchism and the Struggle for the Emancipation of Women* (Bloomington: Indiana University Press, 1991), p. 57.

23. Ibid., p. 115

24. Ibid., p. 116

25. Annabel Rodda, *Women and the Environment* (London: Zed Books Ltd., 1991), p. 111.

26. A number of feminists theorists have explored the false dichotomy between reason and emotion. For a particularly clear and elucidating exploration, see Allison M. Jaggar, "Love and Knowledge: Emotion in Feminist Epistemology," in *Gender/Body/Knowledge*, eds. Alison M. Jaggar and Susan R. Bordo (New Brunswick: Rutgers University Press, 1989).

27. James Baldwin, "Color," in *The Price of the Ticket: Collected Nonfiction 1948-1985* (New York: St. Martins, 1985), p. 320.

PART III

TOWARD A SOCIAL
DESIRE FOR NATURE

CHAPTER FIVE

THE JOY OF LIFE:
THE NATURAL EVOLUTION OF SOCIAL DESIRE

Exploring the social nature of desire has profound implications for understanding the desire for nature. By recasting desire as potentially relational rather than essential, and social rather than individualistic, we are able to rethink not only relationships within society, but society's relationship to nature as well. We see that just as we have the potential to desire social cooperation, we also have the potential to desire an interdependent relationship with the rest of the natural world.

Far from the romantic desire to protect a 'nature' that is a pristine other, a realm prior to and outside of human action, a social desire for nature understands nature as a process of natural evolution in which humanity may potentially play a liberatory role. Departing from a desire for nature that regards human intervention into nature as inherently destructive and 'unnatural', we can begin to consider the 'naturalness' of our social desire to engage creatively in ecological processes. We may begin to see that throughout natural evolution, organisms are marked by a tendency to elaborate upon the natural world through mutualistic activity, creative self-differentiation, and development. These evolutionary tendencies constitute a wider natural history of the social desire for nature.

Social desire does not appear suddenly like a bolt of lightening, or with a wave of a god's finger, but emerges organically through the process of natural evolution. As we have seen, rather than constitute an energy, spirit, or essence that activates an otherwise passive humanity, the socio-erotic is a set of relational potentialities. And unlike ideas of spirit or energy which cannot be explained in organic evolutionary terms, we can indeed, trace the natural history of the socio-erotic.

Thus we have two tasks at hand: to rethink the natural origins of social desire and to cultivate a new social desire for nature. To this, we may add one last task: to develop a way to distinguish between desire that is social and anti-social, rational and irrational. Our discussion of social desire would be meaningless if understandings of what constitutes ethical social desire were left to matters of personal opinion. We must move, then, toward an objective *historical*, rather than personal and relativistic, criterion for distinguishing between social and anti-social desire. To accomplish this task, we might look to the *natural history* of social desire to explore how trends in natural evolution toward mutualism, differentiation, and development may constitute ecological principles that provide a theoretical 'ground' for an objective understanding of social desire.

It is crucial to explore the organic origins and ethical implications of the desire for both social cooperation within society and between society and the natural world. Reflecting upon the origins of this desire within nature itself, we may explore what social ecology has to offer to a discussion of objective criteria for distinguishing between social and anti-social desire, exploring its implications for the desire for nature. Ultimately, we may examine the social desire for nature, moving toward a new revolutionary way to express the yearning for a meaningful and ecological quality of everyday life.

The Eco-Erotic: Principles Of Mutualism, Differentiation, And Development In Nature

To understand the origins of social desire, we may look to natural evolution to find tendencies in nature toward mutualism, differentiation, and development—tendencies that are homologous to dimensions of the socio-erotic. We may call these tendencies in natural evolution the 'eco-erotic' which represents three ecological principles that provide natural evolution with degrees of directionality and stability.

This discussion of natural evolution rests on an understanding of a significant qualitative distinction between the ecological principles I will explore and the dimensions of social desire. While the former exist prior to human history, the latter are inseparable from historical and social constraints that shape and limit the expression of human sociality in all of its forms. It is for this reason that I will not explore sensual or oppositional moments within the natural world. Understandings of sensuality are predicated on a social and historical set of aesthetic, sexual, and relational practices specific to human cultural practices. In turn, the idea of opposition represents a response to social and political institutions created by societies. It is not that I believe that other species are not sensual or that they never oppose obstacles which may confront them. It is that, of the five moments of desire described previously, these two imply greater degrees of subjectivity and consciousness than do the others. To attribute these qualities to species in general in the natural world would run a greater risk of anthropomorphizing.

I have chosen to focus on the three ecological principles of mutualism, differentiation, and development, because they are general and meaningful enough to help illustrate moments of continuity between natural and social expressions of what I am calling the erotic. What I seek here is to establish ecological principles of mutualism, differentiation, and development as prototypical of social expression of desire. I will attempt to show that, while social desire is not reducible to principles that inform natural processes, there does in fact, exist an evolutionary continuity between the socio- and the eco-erotic. Such a discussion hopefully leads the way for a greater appreciation of the 'naturalness' of social desire, that as we shall see, has its roots within a wider natural history.

The Eco-Erotic Principle Of Mutualism

Mutualism is the first principle of the eco-erotic. At the end of the 19th century, social anarchists began to identify mutualistic tendencies in the natural world, tendencies that may be framed in 'erotic' terms. As early as 1891, social anarchist Errico Maletesta challenged social Darwinian and Malthusian theories that portrayed nature as an inevitably competitive struggle for scarce resources, asserting instead that "cooperation has played, and continues to play, a most important role in the development of the organic world."[1] Similarly, Peter Kropotkin began writing about mutual aid in 1890. In his book *Mutual Aid: A Factor of Evolution*, Kropotkin criticized bourgeois theorists for downplaying Darwin's emphasis on the *cooperative* as well as competitive nature of evolution.[2] Kropotkin challenged this interpretation by presenting a series of zoological studies demonstrating examples of inter-species mutual aid as a major factor in species survival:

> ...mutual aid is as much a law of animal life as mutual struggle, but that, as a factor of evolution, it most probably has a far greater importance, inasmuch as it favours the development of such habits and characters as insure the maintenance and further development of the species, together with the greatest amount of welfare and enjoyment of life for the individual, with the least waste of energy.[3]

In addition, Kropotkin regarded the latent sociability of animals as being more than a survival strategy. According to Kropotkin, animals associate with one another because they experience pleasure in so doing, not just because they are obliged to for physical *need* or survival. Identifying a nascent expression of subjective sensuality in the natural world, Kropotkin portrayed the pleasure gleaned from the animal play of higher mammals as a joy of life.

The idea that animal behavior may be driven by something other than utility or necessity, something homologous to human desire, represents a radical break with the Hobbseyan portrayal of nature (and society) as a war of all against all. In turn, by emphasizing the theme of a tendency in the natural world toward mutualism, Kropotkin challenged the Baconian portrayal of nature as an inert passive machine, a portrayal popularized with the emergence of modern Cartesian science. For Kropotkin, a trend toward latent mutualism is constitutive of a development that becomes more complex, rational, and conscious through the evolutionary process. According to Kropotkin, this trend expresses itself "in proportion as we ascend the scale of evolution, growing more and more conscious (eventually losing its) purely physical character...ceasing to be simply instinctive, it becomes reasoned, it becomes a voluntary deviation from habitual moods of life." [4]

Hence, this latent level of mutualistic 'joy' in more simple species gradually gives way to degrees of more intentional and conscious associative 'joy' in more complex species. This trend can be observed in the elaborate grooming behaviors of most primates which serve not only the function of necessary hygiene, but of sensual nurturing and social reassurance as well.

We may trace humanity's social desire for sensual association back to this latent mutualistic tendency in the natural world. The human yearning for a particular quality of association, one that is subjectively pleasurable, finds its origin within latent degrees of subjectivity in the animal world as species strive not only for that which is physically necessary, but that which is qualitatively desirable as well. In this way, we could say that in the natural world, there exist nascent expressions of a sensual and associative subjectivity that become increasingly conscious within humanity in the form of self-conscious sensual and associative desire.

However, we must not overstate the homology between a mutualistic tendency in nature and social desire in society. Kropotkin erred on the side of romanticism in his exuberant anthropomorphic celebration of 'dancing ants'. What makes Kropotkin's discussion radical is not his romantic idealization of animal behavior, but rather, his understanding of the continuity between the tendency for mutualism in nature and culture that runs through natural evolution.

The Eco-Erotic Principle Of Differentiation

Exploring the qualitative and subjective dimensions within species brings us to the second ecological principle of differentiation. The principle of differentiation in nature represents the tendency toward flexibility, spontaneity, and creativity which allows organisms to deviate from established patterns or norms. Geneticist Barbara McClintock, who studied genetic mutations in corn, explored the role of genetic variegation and mutation in natural evolution.[5] In addition to searching for coherent patterns of genetic regularity in corn plants, McClintock explored the seemingly chaotic patterns of corn kernels as well. Out of the differentiated chaos of kernel arrangements, McClintock found larger patterns of regularity, recognizing that the deviations within such patterns were not only inevitable, but developmentally favorable, often leading to vital degrees of organic innovation.

Evolution itself is made possible by the tendency toward innovation that marks both the micro-organic and the organic worlds. In contrast to Freud's belief that Eros creates "the one" out of "the more than one," eco-erotic differentiation tends to make "more than one" out of "the one" through a process of spontaneous complexification. Throughout the evolutionary process, a tendency towards self-differentiation continually opens up new avenues for organic development. Without differentiation, the process of natural evolution would be reduced to mere stasis, repetition, and circularity. It would be what systems theorists call a "closed system" in which organic evolution would never have gotten off the proverbial ground.[6]

Yet the tendency toward spontaneity in nature should not be conflated with an idea of a randomness or incoherence which precludes an organic logic or order. In the same way that unity does not entail the sacrifice of diversity, diversity does not require the breakdown of all coherence or unity. Rather, it is precisely the patterns of regularity within the natural world which lay the ground for creative deviation from those patterns. The developmental process toward ever greater levels of differentiation and complexity takes place within larger patterns that are often marked by dimensions of order, balance, and symmetry. And as McClintock illustrates, out of seemingly chaotic or random genetic deviations within corn plants, emerge new patterns of stability and regularity; patterns that, in turn, serve as the ground for further creative deviation. In this way, there is a dialectical relationship between chaos and order, in which the spontaneous tendency toward disorder is predicated on both a background of order, as well as a reach for new levels of integration and coherence.

Degrees of spontaneity in nature play a crucial and creative role in opening up possibilities for new levels of complex development. This innovative tendency stems from the tendency within nature for organisms to become something else, to complicate things, to make natural evolution innovatively 'messy'. In turn, the breaking of patterns also makes for diverse eco-communities[7] which display a greater chance for survival and sustainability. Ecology has shown that the more complex and diversified a particular eco-community, the greater the chance for species to survive such changes as climatic variation or the introduction of new insects or animal predators. In this way, the deviation or differentiation of particular organisms represents a form of individual flexibility which leads to the greater flexibility of the larger unity.

To return to the homology between the eco- and socio-erotic, we may say that the latent striving in natural evolution for differentiation represents a nascent form of the social desire within individuals and society for differentiation. We could say that humanity incorporates nature's tendency for differentiation, bringing it to a more complex and conscious level of development. The social desire for spontaneous creative expression, for dynamism and change, while not reducible to organic differentiation, resonates historically with this impulse. The spontaneous divergence of corn plants from seemingly stable genetic patterns is evolutionarily homologous to the social desire in society to spontaneously diverge from social and cultural patterns to create something new, to further expand the horizon of freedom, choice, and social complexity.

The Desire For Nature Revisited: Toward A Social Desire For Nature

Yet as in the case of the socio-erotic, the idea of differentiation in the eco-erotic remains unfulfilled unless complemented by the idea of *development*. The tendency toward development in the natural world represents the latent striving for ever greater degrees of *coherent* self-organization and maturation. As explored in the previous chapter, the idea of development is qualitatively different from the idea of simple growth or change. The idea of development implies the movement from that which is more general to that which is more particular, complex, and differentiated. Yet again, within this process of differentiation, degrees of unity and order within previous phases of development are *retained* throughout the process of developmental complexification.

As an organism develops to become something new, it retains its old identity, incorporating and transforming the structure of the old identity into a more mature form. To digress briefly from our discussion of nature, we might consider the experience of meeting an individual we have not seen since they were a child. While we may be struck by their new mature physique, we are often able to identify this new individual precisely because we can derive from this more mature form, a previous less differentiated form. When we say "I know you from third grade!" we are saying that, despite the process of developmental transformation, we understand that the old individual we knew has been incorporated and retained throughout the process of maturation. We are able to see through the more particularized form of the adult presented to us, recognizing the more general form of the child that has been retained.

This cumulative process is integral to an eco-erotic principle of development. It represents the process by which organisms both retain and transform their old identity to become something newer, more complex, and differentiated. The process of 'becoming' that constitutes natural evolution is indeed a process in which organisms 'change' while paradoxically staying the 'same'.

This latent striving toward development in nature represents a process of *self-development* which is not determined, but is endowed with degrees of open-ended and active participation that gradually emerges into the *social* developmental desire in society. In *The Ecology of Freedom*,[8] Bookchin emphasizes the 'self-organizing' properties of organisms, describing the degree to which they actively participate in their own development by shaping and organizing their environment:

> I wish to propose that the evolution of living beings is no passive process, the product of chance conjunctions between random genetic changes...evolution has been marked until very recently by the development of ever more complex species and eco-communities. Diversity may be regarded as a source not only of greater eco-community stability...it may also be regarded in a very fundamental sense as an ever expanding, albeit nascent, source of freedom within nature, a medium for objectively anchoring varying degrees of choice, self-directiveness and participation by life forms in their own evolution.[9]

Bookchin unsettles the idea that organisms merely adapt passively to an already determined environment, asserting instead that organisms become increasingly participatory and self-directed as the evolutionary process unfolds.

Social ecology rejects notions of biological determinism and evolutionary necessity. Moving beyond mechanistic and lawful portrayals of 'nature', it depicts an evolutionary process that is marked by spontaneity and potentiality rather than natural law and pre-determined order. Nature is not a static green box sitting at the edge of society; nor is it a metallic spring that unwinds mechanically. Rather, natural evolution represents the ongoing dance of life itself moving toward ever greater levels of self-expression, eventually giving rise to a potentially rational and desirous second nature. Organisms are marked by a tendency to adapt, modify, and to develop creatively by making nascent evolutionary choices.

This developmental tendency within natural evolution resonates historically with social developmental desire. Just as organisms have a latent desire to fulfill their potential for development, humanity possesses a social desire to develop its unique talents, abilities, and potentialities as well. This striving not only to survive at the minimum, but to cultivate ever new creative ways of relating to the environment is shared by both human and other species on the planet. The tendency toward not only stasis and stability, but toward innovation and development, provides the basis for evolution within both nature and society as well.

However, it is vital to distinguish this social ecological interpretation of ecological tendencies toward mutualism, differentiation, and development from the erotic naturalism of Wilhelm Reich. In the early 1900s, Wilhelm Reich, psychoanalyst and later, physicist, developed an 'erotic' theory of nature.[10] Influenced by Freud, Reich reduced his concept of desir', or *Eros,* to sexual energy. Reich called this sexual energy, *orgone,* demonstrating how it permeated both natural and social worlds. According to Reich, both nature and society are regulated by the basic properties of orgone energy, namely the process of tension and release the same process that marks the sexual act. When this process of tension and release is obstructed, Reich contended, an impairment of biological functioning occurs within the individual organism. Accordingly, Reich identified 'orgonic blockages' as the cause of problems ranging from impaired cellular functioning and sexual/social neurosis to cancer. Throughout his career, Reich advocated for creating a society that would allow for the free flow of orgone energy on both ecological and social levels.

At first view, Reich's orgone energy could be seen as similar to the idea of an eco-erotic. However, there are many differences. First, Reich focused on the *energetic* processes of nature without focusing on the *developmental* process of nature. Fascinated by processes of movement, change, and stimulus-response, Reich was unconcerned with the ways in which such processes differentiated or became more complex through the evolutionary process. Reich expressed a far greater interest in exploring the structural and functional similarities between cells, organisms, and humans than the differences. Accordingly, Reich identified moments of desire in nature which he designated as functionally identical to human desire. For instance, Reich believed that the cytoplasmic movement of cells was functionally identical with the emotional movement or responsiveness of humans. In turn, the expansion and contraction of microorganisms were expressions of pleasure and displeasure that shared functional identity with the emotional correlates in humans. Again, Reich's exclusive emphasis on consistency or functional similarity, rather than differentiated development as well, distinguishes a Reichian from a Bookchinian view of the relationship between organic and social phenomena.

While it is meaningful to explore the similarities between an 'eco-erotic' and a 'socio-erotic', it is also crucial to appreciate that which developmentally distinguishes the two. Whereas Reich was looking for an energetic unity between the desire of all life forms, we need to examine the developmental 'unity in diversity' in which desire itself is engaged in a developmental evolutionary process, moving from moments of organic latency to social and self-conscious actualization.

To reduce the social desire for association, for example, to the ecological desire for mutualism, would be to erase the cultural, social, political, and economic forces that both shape and constrain human associations at any given moment in history. The fact is, the desire to join a worker's collective is not reducible to the mutualism of 'worker bees' that are attracted to a particular hive. Whereas the behavior of worker bees is primarily guided by biological instinct, the behavior of human workers is primarily shaped by self-consciousness and by the social institutions that historically shape notions of work, freedom, and resistance that are fundamental to human history in the modern and post-modern period.

First And Second Nature: A Way To Talk About Evolutionary Difference And Continuity

So far we have discussed the dimensions of the eco-erotic, noting both the evolutionary continuities and discontinuities between the eco- and socio-erotic. In order to further flesh out this discussion, we need to be able to distinguish the eco-erotic from the socio-erotic to demonstrate the differences between the two. Yet if we appeal to conventional categories, we might just assign the eco- and socio-erotic to the categories of society and nature to highlight their differences. We need, then, a way to understand the relationship between ideas of nature and society that will allow us to appreciate the 'evolutionary difference' between the social and natural worlds.

Social ecology differentiates between categories of 'nature' and 'society' revealing a developmental continuum between the social and natural worlds. Referring to two distinct yet continuous phases in natural history, *first* and *second nature*, it illustrates how the latter is derived developmentally from the former. Quite simply, first nature represents all processes and products of natural evolution that emerged from the beginning of the earth's formation through to the gradual appearance of human society. In turn, second nature represents humanity, human consciousness, and human practices including the formation of diverse cultures, the creation of institutionalized human communities, the creation of an effective human technics, the development of a richly symbolic language, and a carefully managed source of nutriment.[11] For example, whereas a tree may represent first nature, a table constructed from that tree represents second nature. In this way, the two categories are not necessarily discrete. With the emergence of second nature, the two 'phases' of first and second nature begin to overlap as human cultural practice informs the processes of first nature. Long before the emergence of capitalism, human societies began to dramatically inform natural processes. Ancient practices ranging from grazing of lands by livestock and hunting and gathering to shifting cultivation and irrigation practices have radically informed ecosystems across the globe for thousands of years. As societies emerged throughout natural history, their practices have always mediated first nature—a mediation that challenges romantic notions of a pure, pristine, or untouched wilderness. Thus, whereas we can differentiate between first and second nature *historically* by identifying two distinct yet continuous phases of natural evolution, it is inaccurate to assert the persistence of two discrete categories once societies begin to emerge within natural history.

We can apply the idea of first and second nature to our understanding of the eco- and socio-erotic. Whereas the eco-erotic represents the tendency toward mutualism, differentiation, and participatory development in first nature, the socio-erotic represents the *social* expression of these desires in second nature. And just as second nature gradually emerges out of first nature through the evolutionary process, the socio-erotic emerges out of the eco-erotic as the latent striving for mutualism, pleasurable creativity, and development becomes increasingly conscious and subjective.

The eco- and socio-erotic represent two major phases within natural evolution. And whereas the eco-erotic of first nature is primarily informed by degrees of biological instinct, the socio-erotic of second nature is primarily informed by cultural practices, social institutions, and degrees of self-conscious choice and intentionality. Thus, the terms first and second nature allow us to point to the evolutionary continuities as well as discontinuities between the eco- and socio-erotic by helping us to see natural evolution as a continuous evolutionary process that is comprised of distinct, increasingly differentiated phases. In this way, the terms 'first' and 'second' nature offer a way to further nuance our discussion of the eco- and socio-erotic by transcending essentialist and dualistic terms such as 'nature' and 'society' .

However, there are those who are concerned that such terms imply a hierarchical relationship between the natural and social phases of evolution.[12] Aware of the ways that ideas of difference have been used to justify the unethical treatment of animals and the destruction of natural processes, many believe that we should, instead, emphasize the similarities between humans and other organisms, asserting that humanity is essentially no different from or even inferior to other organisms.

This sentiment has become popular among many privileged peoples in the era of advanced capitalism. Rightfully dismayed by ecological injustices caused by irrational social relationships constituted by capitalism, patriarchy, racism, and the state, many believe that the cause of ecological destruction is humanity itself, a humanity that has placed itself above nature. In reaction, ecology becomes a form of social criticism that posits nature as everything good that humanity is not. While nature is spiritualized and romanticized, portrayed as a martyred innocent that we must save, the idea of humanity is cast out of notions of earthly paradise that we have constructed in reaction, frustration, and pain. Unfortunately, however, 99% of this humanity, denigrated along with nature, is blamed for the unjust deeds of the 1% in power.

Rather than place the idea of humanity below or above the idea of nature, it is crucial to locate humanity *within* natural evolution in a historical and developmental relationship to other species. While understanding the developmental similarities between humans and other organisms allows us to understand our historical origins and relatedness to previous forms of development, understanding evolutionary difference allows us to understand our unique capacities for innovation, both creative and destructive. For instance, to know our individual potential as human beings, we need to explore what makes us unique, uncovering our particular interests, talents, or desires. To know ourselves, we must cultivate a differentiated sense of ourselves within the context of others, being able to identify that which renders us both like and unlike others.

Similarly, for society to know itself, it must be able to point to that which renders it both like and unlike first nature. To solely emphasize continuities between nature and society is potentially dangerous. For we are then unable to identify both our distinctively liberatory potentialities as well as our harmful capacities. The emergence of society itself represents an undeniable novelty within the whole of natural history. With the emergence of humanity, we see the introduction of novel expressions of abstract language systems, elaborate social institutions, and unique forms of rationality, consciousness, and desire whose liberatory potential has yet to be actualized through the elaboration of a truly humane and ecological society.

It is indeed challenging to find an adequate analogy for the developmental relationship between society and nature. Liberal capitalist society is so thoroughly steeped in the ideology of domination and hierarchy that developmental metaphors often smack of anything ranging from romantic to reactionary. It is then, tempting to merely assert that nature and society are part of one another and leave it at that. But what do we lose when we look only at the developmental continuities between nature and society, ignoring the evolutionary differences? We lose the opportunity to look closely at the organic derivation of society within the natural world, losing too, a vital understanding of our own natural history and distinctive social potentiality. Naming and exploring the developmental relationship between first and second nature allows us to see where we come from, where we are, and what constitutes our unique potential for creating a responsible and ethical ecological future.

The antidote to our negative feelings about humanity (or our 'anti-humanism'), is to see what is best in humanity, tracing the origins of these qualities back to first nature itself, exploring our erotic origins in, and resonance with, natural history. What we like most about the idea of 'nature', the 'innocence' we describe, resonates with humanity's cooperative sensibility—the antithesis to capitalist rationalization, greed, and corruption. What we love most about particular landscapes, the unbounded interplay between symmetry and dissonance, the dance of form, depth, light, and color—these qualities resurface within our own sensual and intellectual creativity. They resurface within our own differentiative desire to combine spontaneity with reason, widening the horizons of meaning, beauty, and poetry. What we savor in 'nature' and society is the expression of the erotic in its many forms: the striving for such relational pleasures as interdependence, creativity, self-determination, and self and collective development—in both the social and natural worlds.

To say that 'humanity' is part of nature means more than acknowledging a biological inheritance from an evolutionary past, more than recognizing humanity's incorporation of ancient cellular structures and spinal columns among the first vertebrates. While appreciating this biological inheritance, we must comprehend the *qualitative* implications of inheriting a biology that is marked by a developmental trend toward increasing complexity and consciousness. We must also recognize that humanity is potentially a qualitative and erotic elaboration within natural history of all that we love about the idea of 'nature': a trend toward increasing sensuality, mutualism, creativity, and the relentless insistence on diverging, ordering, and becoming. Exploring the evolutionary relationship between a first and second nature allows us to understand both the erotic continuity and differences between the natural and social worlds. We may perhaps begin to transcend this anti-humanism, looking back through natural history to see what is best in ourselves winking back at us in a nascent form.

Toward An Objective Understanding Of Social Desire

We may now consider whether there is ethical meaning that can be gleaned from the notion of evolutionary difference. Are there, indeed, ethical implications to be drawn from the fact that natural evolution moves in a developmental trend from the simple to the more complex, from the conscious to the self-conscious, and from the eco-erotic to the socio-erotic? What does it mean that tendencies in first nature toward mutualism, differentiation, and development are part of a larger evolutionary trend toward increasing consciousness, subjectivity, and rationality? What *ethical* sense do we make of Kropotkin's assertion that desire grows more conscious, rational, and voluntary, eventually losing its primarily "physical, instinctive character" as it develops from first to second nature?

The assertion that the potentiality in first nature for mutualism, differentiation, and development becomes increasingly conscious and *rational* as it moves through natural evolution introduces a novel ethical question: If humanity has the ability to consciously respond to its desires, then it has the potential to be *responsible* for its desires as well. That second nature has the potential to mediate, reflect upon its impulses, inclinations and yearnings, implies an evolutionarily unique expression of 'desire'. Far from the Freudian view of desire as primarily pre-rational and animalistic, an impulse that must be suppressed by the rational ego, we may now appreciate the rational and social dimensions of desire itself. Unlike non-human species whose latent subjectivity is highly mediated by biological instinct, humanity's biological instincts are largely mediated by consciousness, rationality, and history. We are, after all social creatures whose desires are informed, for better or for worse, by the idiosyncrasies of the particular cultures in which we live.

The fact that humanity can reflect upon, choose, and even *institutionalize* which shades of desire to act upon introduces an ethical dimension to the idea of desire. For now we are obliged to ask: what kind of desire *ought* humanity to express? Is it is equally valid to express an individualistic desire that inhibits others from fulfilling their potential for freedom as it is to express a social desire that enhances the subjectivity of others? Would we assert that it is as valid to destroy eco-communities to fulfill individualistic yearnings for power and profit as it is to enhance ecological complexity for the good of all? What criteria do we use to evaluate the validity of our social desire? When we express the oppositional desire to transform social and ecological reality, how do we distinguish between reconstructive activity that is rational from irrational, social from anti-social, or erotic from anti-erotic? As I have suggested earlier, the 'desire for nature' represents a social construct that may be expressed in wide range of forms. For some, a romantic, anti-humanistic desire to annihilate the human species to protect the natural world is a valid desire for nature. For others, such as the CEOs of Novartis, a capitalist desire to reduce the biological complexity of Amazon rain forests to 'cell lines' to be patented and sold represents a valid way to desire 'nature'. Still, for others, the desire to create directly democratic institutions to empower citizens to engage creatively and cooperatively with natural processes, represents an ethical 'desire for nature'.

In light of this ideological 'diversity', who, indeed, is to say which 'desire for nature' is objectively more rational, ethical, or valid than any other? Why shouldn't the privileged express their sensual desire for nature, relaxing at lush island resorts where indigenous workers refill their Margaritas? Why shouldn't white middle-class Americans express associative desire by communing with 'nature' by appropriating Native American rituals while actual native peoples can no longer practice such rituals because their lands are stolen or poisoned by toxic waste? Why shouldn't privileged First World theorists express their differentiative desire for nature by writing elaborate theories that blame immigrants and women for destroying ecosystems by 'overpopulating'?

In turn, can we assert an objectively rational ground for a social desire in general? While we have the potential to cultivate the socio-erotic in a cooperative direction, we also have the capacity to direct our desires in an authoritarian or capitalist direction, using sexuality for domination and intimidation and creativity for profit to enhance personal status and authority. If we fail to identify a set of criteria for making such distinctions, we have no way of asserting that the social desire for non-hierarchy is more ethical or 'erotic' than the desire to construct hierarchy, or that ecological cultivation is more ethical than a capitalist rationalization of nature. Without a stable, general, or objective criteria for determining what makes social desire more ethically valid than anti-social desire, the quality of our relationships with each other and with the rest of the natural world becomes just a matter of arbitrary personal opinion.

To transcend this relativism, we must anchor ideas about the 'desire for nature' in something more stable than subjective inclination. The real question becomes: on what can we ground an organic rationality that will be able to distinguish between desirous actions that enhance or threaten an evolutionary trend toward increasing social and ecological complexity?

Natural Evolution As A Ground For A Social Ethics Of Desire

To address this problem of objectivity, we might again turn to the natural philosophy of social ecology. *Dialectical naturalism* is an approach to natural philosophy developed by Bookchin which builds on, yet transcends, the dialectical traditions of such thinkers as Hegel and Marx.[13] For Bookchin, 'nature' is a dialectical process of unfolding that is marked by tendencies toward ever greater levels of differentiation, consciousness, and freedom. While it is beyond the scope of this book to fully explore this rich and important theory, we may look briefly at a few key concepts drawn from Bookchin's dialectical naturalism to elaborate our understanding of social desire.

Bookchin appeals to the idea of natural evolution to establish ecological principles which we may be utilized to evaluate the ethical dimensions of our social desire. As we begin to understand 'nature' as a process of natural evolution, we recognize the ethical implications of the idea of 'nature' as flowing out of the idea of evolution itself.

Locating humanity within natural evolution raises an ethical question: what is humanity's *role* within the process of natural evolution? If humanity has the potential to build upon this evolutionary trend toward complexity, *ought* it to do so? Again, we might ask, is it equally rational for societies to reverse this evolutionary trend by institutionalizing hierarchical social relationships based on command and control, while also undoing horizons of biological and cultural differentiation or diversity? In turn, is it equally 'rational' for humanity to reverse the developmental directionality of natural evolution, a trend that has led from simple unicellular organisms to increasingly complex species, from consciousness to self-consciousness, from simple to more complex expressions of subjectivity? As social ecology illustrates, this reversal is irrational for it contradicts the *developmental logic* of natural evolution itself.

The ecological principles of mutualism, differentiation, and development provide a set of criteria by which to measure the ethical validity of human action. Again, as social ecology shows, humanity *ought* to further this trend toward increasing mutualism, differentiation, and development and that, in contrast, it is *irrational* to counter this evolutionary trend. We may assert that the social desire to create cooperative institutions and social practices is ethical and rational because such practices further the trend in natural evolution toward ever greater levels of mutualism, differentiation, and developmental complexity that provide the basis for natural evolution itself.

For instance, the practice of direct democracy requires and enhances degrees of mutualism, differentiation, and development more than does the practice of representational democracy. Direct democracy is a process in which members of a local community are empowered to participate directly in creating the public policy that gives shape to their everyday lives, both public and private. Unlike a representational democracy in which citizens elect a centralized body of 'politicians' who make decisions on their behalf, a direct democracy is one in which decision-making power is decentralized among citizens themselves.[14]

A direct democracy supports the principle of mutualism or non-hierarchy by creating a forum in which an entire community is engaged in participating cooperatively to discuss, debate, and determine public policies. Direct democracy draws from the principle of differentiation or social complexity by encouraging a rich process of pubic discussion in which a diversity of perspectives are presented and considered. Difference of opinion is welcomed as members of a community continually work to nuance and complexify their understanding of freedom. The process of self-reflection, the give and take of dialogue, the intricate mediations of self-consciousness and consideration for others, requires and nurtures a highly differentiated 'body politic', a body of citizens capable of thinking for themselves. In turn, the idea of direct democracy draws from the principle of development by encouraging members to cultivate their abilities to discuss and debate with others in a collaborative decision making process. Through the process of participating in a direct democracy, members develop both the capacity for self-knowledge and the maturity to critically consider the perspectives of others as well.

In contrast, a representational democracy (really, a contradiction of terms) reduces dimensions of mutualism, social complexity, and development. Countering the principal of mutualism, a representational democracy reduces citizens to individual voters or separate 'constituencies' who back particular representatives, depriving them of the opportunity to work cooperatively to make policy that provides for a common good. While direct democracy offers a rich process of discussion and debate that engages a wide range of complex social issues, a representational democracy opposes the principle of social differentiation by reducing social issues to campaign slogans and 'platforms' which simplify social and political issues to appeal to the lowest common denominator. Finally, a representational democracy goes against the principle of development by centralizing not only decision-making power, but also by depriving citizens the opportunity to develop their abilities to think, speak, write, and debate about public issues that determine their very own lives.

We can apply the same principles to a discussion of economics as well. The practice of a directly democratic economy, or a moral economy fosters social complexity.[15] According to Bookchin, a moral economy is based on the principle of mutualism as goods are produced and distributed democratically according to needs and abilities of all members of a community. Fostering relationships based on inter-dependence and complementarity, a moral economy allows communities to try to minimize, rather than enhance, disparities of wealth or privilege that could otherwise emerge from physical differences and abilities. The practice of complementing individual need with the abilities of the community allows for ever greater degrees of participation, freedom, choice, and subjectivity by all, for all.

A moral economy is in accordance with the principle of social differentiation and complexity as community members reflect upon, discuss, and decide how to provide for a common good. The rich social relationships that emerge as community members provide collectively for their own needs and desires opens up ever new avenues for the development of creativity, self-determination, and cooperation. Free of the constraints of a market economy that requires workers to stunt their own development by spending the majority of their lives engaged in alienated labor, a cooperative moral economy supports the principle of development by freeing people to pursue a range of creative and intellectual developmental desires.

In contrast, a capitalist market economy reduces mutualism, social differentiation, and development. Based on social relationships of owner/worker and consumer/producer, capitalism counters principles of mutualism and differentiation, supporting instead a simple system of command and control. For example, within an increasingly 'global' capitalist economy, a handful of transnational corporations autocratically determine what shall be produced, by whom, and at what cost for people and eco-communities throughout the world. Rather than local communities participating in a decentralized way, determining their own needs and desires in a spirit of mutualism and social complexity, the corporation determines, through market research and media manipulation, what 'consumers' will buy, centralizing the power and resources that determine the social and ecological fate of the many. Capitalism counters the principle of development by reducing members of a community to 'consumers' and 'workers' whose labor and lives are marked by degrees of alienation. Deprived of the ability to develop rich social and ecological networks based on inter-dependence and mutual aid, people are reduced to buyer and seller as the natural world is stripped and sold, reversing the developmental trend toward biological complexity.

Having looked briefly at the examples above, we may now assert that it is *objectively true* that the social relationships surrounding participatory democracy and a moral economy are more likely to enhance the evolutionary tendencies toward mutualism, differentiation, and development than are the social relationships surrounding a state-run democracy and a capitalist economy. And when we say that it is objectively true, we mean that it is not relative, arbitrary, or a matter of personal opinion.

If as we have shown, 'nature' is a natural history, a process of organic development marked by a trend toward increasing complexity and freedom, then a social desire for nature implies a desire to play a creative role in furthering this trend. It is indeed irrational to reverse the natural and social complexity that has emerged throughout natural history. It is 'irrational' for those in power to make most of the earth's population unfree, to simplify social relationships to 'top-down' and 'command and control' characteristic of centralized and hierarchical structures. It is irrational to lull individuals and communities into mass conformity and expedience, coercing them to embrace a simple 'blind faith', or an 'unquestionable authority'. Finally, it is irrational to 'undo' the rich complexity of social and eco-communities that evolved over thousands of years, giving way to degrees of increasing flexibility, creativity, stability, and complexity.

In contrast, it is organically rational to elaborate upon this evolutionary trend—to organically 'complicate', rather than simplify, social and ecological reality by creating institutions that allow people to be freer, more joyous, and creative.

Organic Objectivity: A Ground That Moves

Yet here we witness a new approach to questions of objectivity. The objective dimension within social ecology's ethics, far from being rooted in deterministic universal 'natural facts', is rooted in the idea of general, nascent, and organic *potentiality*. Here, the understanding of 'objectivity' represents a recognition of an identifiable, stable, yet dynamic trend toward the potential for increasing complexity and freedom in natural history. The 'ground' for this 'organic objectivity' is paradoxically 'unstable'—it is, as social ecologist Amy Harmon says, a "ground that moves."[16] Rather than be anchored in static biological facts, it is anchored in the 'flexible' field of potentiality that allows for ever greater degrees of stability and order to emerge within the process of natural evolution.

Again, such socio- and eco-erotic principles of mutualism, differentiation, and development are not reductive, essential, or deterministic 'natural facts'. Instead, they are complex and rational organizing *tendencies* that give shape, symmetry, and directionality to the process of natural evolution that are open-ended, diverse, and multi-directional, rather than determined or unilinear.

As a non-deterministic perspective, social ecology does not view this trend toward increasing mutualism, differentiation, and development, as the 'dominant' trend in natural or social history; nor does it propose that this trend will necessarily triumph over the *irrational* anti-social tendency toward social hierarchy, homogenization, and simplification. For the fact that particular societies today are characterized by irrational and tenacious forms of hierarchy that reduce social complexity and interdependence and that global capitalism is currently 'undoing' the process of natural evolution by simplifying the environment, is testament to the unactualized potential of societies to participate creatively and rationally in elaborating the evolutionary process.

The trend toward a social desire based on ecological principles of mutualism, differentiation, and development, while not the most pervasive trend, is 'objectively' the most *promising* and *rational* trend, both ethically and politically. For, when societies elaborate upon such a trend, they open the way for greater evolutionary choice and social freedom. It is on this basis that we may ground an ethics of social desire on something more stable than relative or arbitrary 'personal opinion'. The decision to actualize our social desire for mutualism, differentiation, and self-organized development, represents an organically rational expression of desire, for it allows us to participate in elaborating upon, rather than reversing, the evolutionary process itself.

The Desire For Nature Revisited: Toward A Social Desire For Nature

If the rational expression of social desire strives to enhance social complexity, then a rational social desire for nature would strive to enhance ecological complexity as well. Instead of idealizing and preserving 'pure' peoples, times, and places, a social desire for nature leads us to contribute to the diverse and interdependent splendor of eco-communities, elaborating upon the subjectivity in first nature by engaging in practices that enrich biodiversity, stability, and complexity.

Exploring a social desire for nature offers a way to draw meaning out of our sensual, associative, differentiative, and developmental relationships to the natural world. It allows us to point to what is meaningful in the idea of nature without relying upon reductive notions of spiritus, energy or natural essences. In an era in which social relations to nature are reduced to capitalist commodification, we need a way to point to those aspects of our relationship to natural processes that cannot and should not be reduced to relationships of profit and production. Moving away from a language of capitalist rationalization, we need a way to describe the qualitative dimension of our relationships to the natural world that are sensual, cooperative, creative, and elaborative.

Five Dimensions Of The Social Desire For Nature

To begin, a sensual desire for nature is the yearning to taste, touch, smell, hear, and see the creative magnificence of the natural world. Unlike a romantic sensual desire predicated on a people-free 'natural purity', a social-sensual desire for nature appreciates what Donna Haraway refers to as a 'cyborgian' interplay between human technics and the natural world.[17] In this way, a social-sensual desire for nature is non-essentialist, a craving not for pure essences of a bounded idea of 'nature', but instead, a delight in the delicate phasing of natural evolution into the social.

When we stand upon a mountain, looking out, savoring the elegant expanse of forest and plain, instead of relishing in the absence of humanity in the vista, we may recognize our place in the scene, appreciating our potential to glean sensual, philosophical, and aesthetic meaning from this evolutionary process that unfolds before us. Any moment of desiring or loving the sensual qualities of 'nature' is a deeply social act, located within social history as well as within a wider natural evolution.

In addition, a social-sensual desire for nature entails stretching conventional understandings of what constitutes a 'nature' worthy of appreciation. Moving beyond romantic understandings of 'nature' cast within the idioms of the rural and the wild, we may include the cityscape as an expression of natural evolution as well. Although the city has been reduced to a dense population clustered around centers of industrial capitalism, even within these centers, there exists the sensual yearning for clean tree-lined streets, city parks, open-air cafes, community gardens, and farmers' markets. By expanding our notions of 'nature' to include cities, we include the urban within discussions of quality of everyday life, appreciating the places where much of the world's population lives, struggles, and despite it all, often thrives.

In turn, an associative desire for nature incorporates this sensual appreciation for natural processes, transforming it to a sense of association with the natural world around us.

An associative desire for nature, often referred to as feeling 'at one with nature', represents our joy in empathizing with other species, identifying with the larger process of natural evolution that binds each of us to every organism on the planet. Yet, again, in contrast to a romantic associative desire for nature, a social-associative desire extends this empathy to the rest of humanity, wanting not to transcend our humanity to love a 'pure nature', but to join with the rest of humanity to create a world that is ecologically whole. Feeling at one with nature means feeling solidarity with communities who have emerged from and dwell within the places that we love; it means becoming allies in the twin struggles for social and ecological justice.

A differentiative desire for nature means that while feeling at one with nature, we understand this oneness to represent a unity in diversity. It means that we can hold the sense of being both similar to, and distinct from, other species. While retaining the sensual and associative desire to be part of the natural world, we can complement this yearning with the striving to understand that which makes humanity evolutionarily distinct. Thus, while standing on top of a building or mountain peak, we can include ourselves within the picture. We can understand that we are both similar to and different from the other organisms that slither, crawl, and fly through the sensual field.

We express our *creative* differentiative desire for nature when we draw meaning from our relationship to the natural world. Creative differentiative desire for nature entails the desire to highlight the poignancy of particular moments of natural evolution by representing the earth's beauty through such mediums as philosophy, poetry, song, dance, or painting. As this desire is highly culturally mediated, its expression reflects the values and practices of particular peoples. For some cultures, it entails differentiating natural processes through a fluency in such scientific practices as biology, physics, or ecology; for others, it entails the creation of practices of herbal medicine.

The desire to give names to places and species, represents our yearning to translate the natural world into terms we can relate to, order, and know. Unlike a capitalist desire to taxonomize species for the sake of control and profit, a social desire seeks to name and distinguish species for the sake of knowledge, pleasure, and ecological enhancement. Creative differentiative desire is the yearning to sensitize ourselves to our relationship to the natural world, to draw philosophical and aesthetic meaning from the patterns, symmetries, and rhythms that continually unfold around us.

In this vein, a developmental desire for nature entails wanting not only to know, or differentiate particular 'moments' of natural evolution, but to actively participate in this development in a complementary fashion, using ecological technologies, art, language, and other social practices to elaborate upon the trend toward diversity, complexity, and subjectivity. We may express this developmental desire through creating ecological practices such as solar, wind, and water power, or by practicing organic agriculture and edible landscaping to enrich the eco-communities in which we live.

In turn, we express our developmental desire for nature not only by expanding the richness of the biological horizons around us, but by expanding our consciousness as well. As natural evolution represents a trend toward increasing subjectivity, humanity has the potential to further expand the horizons of consciousness, by elaborating upon the idea of freedom itself. Throughout history, emerging in tandem with the emergence of hierarchy, surfaces the idea of freedom. Each act of writing, discussing, debating, or theorizing about freedom constitutes an expression of the developmental desire to widen the horizon of what we can know and think about what it means to live with liberty and integrity. By striving to further differentiate ideas of freedom, we bring human consciousness, evolving for thousands of years, to new levels of complexity.

An Oppositional Desire For Nature: Toward An Ecological Politics

The current ecological crisis serves as a bitter reminder that our social desire for nature must be translated into political action. It would be naive to believe that a simple 'paradigm shift' to a new set of understandings about nature and desire could abolish social and ecological injustice. For flowing through and around such understandings are social *institutions* of capitalism, the state, racism, and patriarchy which shape particular ways that we relate to the natural world as well as to each other. We need, then, to cultivate an oppositional desire for nature, a rational yearning to oppose all institutions and ideologies that are reversing the trend toward natural evolution by destroying biological and cultural diversity and inter-dependence across the planet.

As previously discussed, there are three moments to oppositional desire. In the first *critical* moment, we begin to analyze social relationships or institutions, assessing the extent to which they enhance or reverse the trend in natural evolution toward increasing mutualism, differentiation, and development. Here, for instance, we look critically at social relationships such as the state and capitalism that inhibit full and direct participation of citizens, reducing them to passive consumers of pre-packaged representatives. We look as well at capitalist activity that hoards native lands, disenfranchising diverse cultures into extinction, and driving species into extinction through pollution and eco-system destruction.

In the next phase of oppositional desire, the moment of *resistance*, we begin to resist these institutions, protesting specific harms that they cause, while popularizing a general critique of the implications of their hierarchical structure. A resistant dimension of oppositional desire for nature is expressed by environmental groups who link the general problem of capitalism and the state to particular moments of ecological destruction. For instance, during the campaign against Hydro Quebec, spokeswoman Winona La Duke contested the building of a system of dams at James Bay that would flood thousands of acres of native land in Canada and the U.S., identifying both capital and state structures as playing a crucial and devastating role in social and ecological devastation. The oppositional desire expressed by indigenous peoples, feminists, social anarchists, and social ecologists—all those fighting for social and ecological justice—represents moments of resistance against the qualitatively dangerous aspects of the hierarchical structure of the state and capital.

Finally, oppositional desire would be incomplete if it were not fulfilled by a *reconstructive* moment. For the struggle for freedom assumes two forms: while 'negative freedom' represents the desire to negate, or abolish unjust institutions, 'substantive freedom' is the assertion of that which must replace those negated structures. Again, while negative freedom is a demand for 'freedom from' particular forms of injustice, substantive freedom is a demand for the 'freedom to' create new institutions that will improve the quality of life for all.

And so, as we move into the reconstructive moment of oppositional desire, the moment in which we consider our substantive desires, we now face a series of intriguing questions: what quality of social relationships is rational to desire? What kinds of social relationships will allow us to further the evolutionary trends toward social and biological complexity and freedom? And what kind of *political* institutions will best facilitate the fulfillment of rational social desire? Perhaps most important, we need to think about what *objective criteria* we may use to determine what constitutes social relationships that are conducive to creating a socially and ecologically just society. The answers to these questions represent the core of revolutionary praxis, and clearly, cannot be sufficiently explored within the scope of this book. However, we may take a brief look at some key issues that we must consider as we begin to approach such questions of social and ecological reconstruction.

To begin with a caveat, it is crucial to emphasize that such questions should not lead to a series of static formulas that dictate how to 'universally' engage in creating a new ecological politics. The revolutionary process, the movement from where society is to where it ought to go, must be created by the very people who are engaged in particular struggles for freedom. However, by invoking an organic rationality, we may explore how particular communities, in concert with other communities, may think about how to develop a set of political practices that are meaningful and relevant to their own needs and desires.

For example, the principle of mutualism may serve as an objective criterion to which different communities can appeal as they think about how to create rational political practices. In this dialectical process, communities may both differentiate and retain the general principle of mutualism to create new forms of non-hierarchical self-government. For instance, the idea of a New England town meeting represents a 'differentiated' form of the general idea of 'mutualism'. Although the idea of a New England town meetings is not reducible to the general idea of 'mutualism', the general idea of mutualism is dialectically *retained* within the particular idea of a New England 'town meeting'. Again, the idea of a New England town meeting is a political practice developed by a particular group of people at a specific time and place within history. Yet, within this particular historical institution is the general idea of 'mutualism' that existed centuries before the New England town meeting ever came into being.

By thinking about how to particularize a general and objective organizing principle such as 'mutualism', communities may begin to think rationally about how to create ecological political structures. It is through the dialectic of public debate and discussion that citizens move from the general to the particular, differentiating such nascent ideas as mutualism into a multitude of public policies that will shape political, social, and ecological practices within a particular community.

Yet once again, these principles do not represent deterministic 'natural facts'. Rather, they constitute nascent yet stable objective *potentialities* that may be worked like clay by citizens as they respond to the particular sensibility, culture, and ecology of their own community. They represent potentialities that have been actualized throughout the evolutionary record, giving rise to a world that is increasingly complex, diverse, conscious, and free.

A social desire for nature requires the reclamation of direct political expression by local people. Only by participating in a state-less direct democracy, will people to begin to articulate the grounds for a new non-hierarchical ecological society. However, localism alone is insufficient for creating the broadest context for democracy. The principle of 'localism' represents only one moment within a dialectic between unity and diversity. A world of 'diverse' self-governing localities could mean a string of parochial islands empowered to 'do their own thing' without being accountable to a wider community. Thus, this spirit of diversity needs to be complemented by a unifying trend as well. A confederation is necessary that would bring together a community of self-governing local cities, towns, and villages who are united through a common commitment to general principles of cooperation and non-hierarchy   (mutualism),   self-determination   and   participation, (differentiation), and development.[18]

A rational desire for nature, then, would lead us to establish cooperative political institutions in which we could create a free and ecological society. It would move us to foster complex and social relationships both within society and with the rest of the natural world in order to build upon the objectively identifiable trend in natural evolution toward mutualism, differentiation, and development.

We must cease identifying such abstractions as 'humanity', 'technology', or 'industrial society' as the cause of ecological problems, or goading 'Third World women' or 'immigrants' for causing ecological harm. It is time to begin to critique the *social relationships* in society, particularly those that constitute our systems of government and economics, understanding their role in perpetuating ecological injustice. An oppositional desire for nature moves us to create a new kind of society in which we are empowered to determine our *social* relations with nature.

The social desire for nature is worlds away from a romantic or spiritual desire to protect an ecological 'purity' or 'integrity' that is 'above' us, or 'in' us. Rather than regard nature from afar, starry-eyed, and yearning, we may recognize ourselves as part of a developmental and creative process called natural history, recognizing in turn, creative and liberatory potentialities within ourselves. After all, we are each organically derived from first nature, we each represent a distinctive 'moment' within a larger natural history.

As we begin to understand our own natural history, we may begin to reinterpret the potentialities latent within our own second nature. By understanding the tendency within first nature for mutualism, differentiation, and development, we are able to comprehend the organic dimension of our own desire for non-hierarchy, creativity, development, and ultimately, freedom. A social desire for nature moves us to regain the courage to see what is best in ourselves, to appreciate our potential to create social and political institutions that bring out what is most empathetic and intelligent within humanity.

The revolutionary impulse is fiercely organic, traceable to the impulse toward creativity and development in first nature itself. Yet all over the planet, the trend toward increasing mutualism, differentiation, and development which has been evolving since the beginning of natural history, is at risk of being reversed completely. By recognizing the very tenuousness of natural evolution, we see that we can no longer reify nature as a kind of spirit, eternal flame, or energy which remains the same while enduring throughout time. Unlike an enduring spirit, natural evolution is a process which can either move forward or regress into simpler phases. Each time a wetland is 'filled in' and a shopping mall shoots up, we lose one more horizon in which both first and second natures are able to unfold.

Creating the conditions for social and ecological complexity is not only evolutionary, it is revolutionary. By creating social and political institutions which encourage first and second nature to express what is most creative and cooperative, we create erotic resonance between natural and social phases of natural history. It is then that the socio-erotic and the eco-erotic meet: in the work of creating a socially and ecologically desirable world.

Notes

1. Errico Malatesta, *Anarchy* (Great Britain: Freedom Press, 1974), p. 26.

2 Peter Kropotkin, *Mutual Aid: A Factor of Evolution* (Boston: Extending Horizons Books).

3. Ibid., p. 6

4. Ibid., p. 6.

5. Evelyn Fox Keller, *A Feeling for the Organism: The Life and Work of Barbara McClintock* (New York: W.H. Freeman and Company, 1983), p. 97.

6. Bookchin offers a convincing critique of the limitations of systems theory in "Toward a Philosophy of Nature" and "Thinking Ecologically" in *The Philosophy of Social Ecology* (Montreal: Black Rose Books, 1995).

7. Seeking to avoid a mechanistic 'systems' language, Bookchin prefers the term 'eco-community' rather than eco-system, emphasizing the relational and holistic qualities of natural processes.

8. Murray Bookchin, *The Ecology of Freedom* (Palo Alto: Cheshire Books, 1982).

9. Ibid., p. 54.

10. Reich, as a post-Freudian Marxist, sought to create a totalizing theory of human behavior that would have revolutionary implications. For a broad overview of Reich's work, see *Wilhelm Reich: Selected Writings, an Introduction to Orgonomy* (New York: Farrar, Straus and Giroux, second printing, 1974).

11. Bookchin, *The Philosophy of Social Ecology*, p. 119.

12. Radical ecologists reservations regarding the discussion of first and second nature reflect deeper concerns regarding "anthropocentrism" in general. For a provocative exploration of and response to these concerns, see Bookchin, *Re-Enchanting Humanity: A Defense of the Human Spirit Against Anti-Humanism, Misanthropy, Mysticism and Primitivism* (London: Cassell, 1995).

13. For a more elaborate discussion of dialectical naturalism, see Murray Bookchin, "Thinking Ecologically," in *The Philosophy of Social Ecology: Essays on Dialectical Naturalism* (New York: Black Rose Books, 1990).

14. The idea of direct democracy is directly tied to Bookchin's theory of libertarian municipalism which entails building a confederation of municipalities engaged in a process of direct-democracy. For an introduction to the theoretical ground for libertarian municipalism, see Janet Biehl, *The Politics of Social Ecology: Libertarian Municipalism* (Montreal: Black Rose Books, 1998).

15. See Murray Bookchin, "Market Economy or Moral Economy?" in *The Modern Crisis* (Philadelphia: New Society Publishers, 1986).

16. I thank Amy Harmon for this phrase that she coined during a class at the Institute for Social Ecology, summer 1997.

17. See Donna J. Haraway, "A Cyborg Manifesto," in *Simians, Cyborgs, and Women: the Reinvention of Nature* (New York: Routledge, 1991).

18. Murray Bookchin. Lecture at the Institute for Social Ecology. Summer 1996. For a wider discussion of confederalism, see Janet Biehl, *The Politics of Social Ecology*.

CHAPTER SIX

ILLUSTRATIVE OPPOSITION: DRAWING THE REVOLUTIONARY OUT OF THE ECOLOGICAL

The project of incorporating a broad revolutionary analysis into particular struggles for ecological justice can be daunting. Each night the news presents us with yet another immediate ecological crisis that demands our attention. Confronted with stories of greenhouse-related disasters, environmentally induced illnesses, or rising levels of pollution, we feel overwhelmed as we try to prioritize our ecological agendas, attempting in turn to link particular struggles for ecological justice to questions of deeper political change. We want to go beyond pragmatic environmentalists who focus on single-issue reforms, yet we are faced with a dilemma: While we know it is crucial to engage in particular ecological struggles, while we know that such struggles are necessary to slow down the pace of wider ecological collapse, we also know that addressing single issues alone is insufficient to bring about radical social and political transformation. We need, then, to explore ways to engage in particular, necessary ecological struggles while drawing out a sufficient revolutionary vision for a new desirable ecological society.

Necessary vs Sufficient Conditions For Political Transformation

Movements for social or ecological change focus primarily on that which is *necessary* to remake society. Whereas many in the Old Left regarded the abolishment of material inequity to be the most necessary condition for a free society, in the 1970s, radical feminists asserted that social justice would necessarily be won with the transcendence of patriarchy. Similarly, many involved in the Civil Rights movement of the sixties believed the elimination of racism to be a primary necessity around which wider social change would unfold. For many in these movements, the abolishment of one specific form of hierarchy was viewed as necessary for radical social transformation. In such movements, people often reasoned: "Once we dismantle this form of hierarchy, other forms will dissolve as well." In this way, what is necessary was conflated with what is sufficient. And still today, we often believe that if we succeed in the necessary task of abolishing one specific form of hierarchy, then this necessary act will be *sufficient* to create a free society.

What is necessary is not the same as what is sufficient. For instance, if we want to boil water, we need to fulfill a few necessary conditions: water and a heat factor which can raise the temperature of the water to 212 degrees. We recognize that if we have only one of the necessary conditions, a pot of water for example, it alone will represent an insufficient condition for boiling water. In the same way, if we have only a heating coil raised to 212 degrees with no water present, the heating coil will represent an insufficient condition as well. Or, if we have a pot of water at one end of a room and the heating coil raised to 212 degrees at the other end of the room, we will still lack the sufficient condition for boiling water—even though we have organized the necessary conditions for boiling water to occur at the same time. If we think only in terms of what is necessary, we may spend hours staring bewilderedly at a pot of unheated water, or at a heating coil, or we may move the heating coil and the pot of water around the room, wondering why we are unable to make the water boil.

Obviously, most people do not have to think critically about the necessary and sufficient conditions for such everyday activities as boiling water. We know intuitively and rationally through conventional logic that the sufficient condition for boiling water represents the accumulation of the necessary conditions for boiling water (water and a heat factor), arranged in a particular physical and temporal relationship to each other. In this way, we understand implicitly that the sufficient condition represents a holistic, accumulative, and integrative whole comprised of all necessary conditions for making water boil.

However, we run up against the limitations of the boiling water analogy when we begin to think about the necessary and sufficient conditions for social and ecological change. For while the conditions that allow one to possess a pot and a heating coil might be clearly social and arbitrary, the mechanics of boiling water dwell largely within a world of physical, inorganic processes that pertain to the movement of heated water molecules. Such an event can occur independently of human action, as in the case of a forest fire boiling ground moisture into wisps of steam. In contrast, the event of revolution is a distinctly *social* phenomenon existing within the realm of potential freedom rather than natural law or necessity. And while this inorganic analogy is in itself insufficient for providing us with a plan to create a revolution, we may use this analogy to begin to think through the necessary and sufficient conditions for an ecological and social revolution. We may ask ourselves: What are the necessary and sufficient conditions to "heat up" society to produce a revolutionary situation?

As we think through the necessary and sufficient conditions for social and political change, the sufficient condition must be understood to be just that—*sufficient*—neither perfect, nor a determined end in itself, but an incomplete beginning. Hence, the sufficient condition is not a deterministic factor. Just because we may have the pot, the water, the heating coil, the right time, and the right place, a great rainbow could majestically appear outside the window and we could find ourselves wholly disenchanted with the idea of boiling water after all. Or the pot could turn out to have an undetectable leak. The sufficient condition means merely that we have fleshed out the idea of necessity *enough* to begin the work that is set out before us. It does not mean that we will be successful in our work, or that the work will turn out to be what we had in mind. It only means that we have a good enough chance, that we have done almost all that we can to increase the likelihood that we will actualize our goals. The sufficient condition, then, represents a glorious point of departure, open-ended as the utopian horizon whose band of brilliant color recedes incrementally as we make our approach so that we never arrive but forever enjoy the desirous and sensuous apprehension of arrival.

In embarking upon this question, we see, as already stated above, that most movements for social change conflate that which is necessary with that which is sufficient. People often select a single issue, source of oppression, or form of hierarchy as the sole focus for necessary social action, never thinking through the sufficient condition for a free and ecological society. However, when we begin to think holistically, we begin to see that the sufficient condition for an ecological society represents the accumulative integration of non-hierarchical institutions and an ecological technics, ethics, and sensibility.

As social anarchism implies, unless we abolish *hierarchy in general* as the sufficient condition for a free society, specific forms of hierarchy may endure. The idea of abolishing only specific forms of hierarchy (such as the State, capitalism, racism, and patriarchy), while *necessary*, proves over and over again throughout history to be woefully insufficient. For instance, while Marxian socialism seeks to abolish hierarchies derived from material inequities, hierarchies such as the State and patriarchal institutions remain largely unchallenged. Similarly, while liberal feminists seek to abolish hierarchies that exclude women from male dominated social and political institutions, hierarchical structures such as capitalism remain unchallenged, leaving women as well as men to be exploited by capitalist production. What is more, while it is necessary to eliminate specific forms of hierarchy such as capitalism and women's oppression, this elimination is not only insufficient for creating a new world, it is even *compatible* with the survival of many other forms of hierarchy.

The compatibility between non-hierarchy and hierarchy can be quite insidious. If we were to eliminate racism, if we were to create the social conditions in which people of all ethnicities were treated equally, capitalists and the State could still refine other criteria such as age, sex, or class, by which to justify social exploitation. In this way, hierarchical systems such as capitalism and the State are compatible with the non-hierarchical conditions of ethnic equality. Or in the event of a non-sexist society, there could conceivably coexist a capitalist and statist society that bases privilege primarily on class and race, rather than on sex. A society organized around egalitarian sexual relations is potentially compatible with a racist, classist, and statist society. What is more, we could conceivably eliminate the idea of dominating nature, establishing a social 'reverence' for the natural world such as expressed by ancient Egyptians, Mayans (or Nazis, for that matter), while still maintaining immiserating social hierarchies. Finally, we could even imagine dismantling the hierarchy of the State only to find that hierarchical corporations take over the management of social and political life completely.

Hierarchy is much like a cancer which, if not rooted out completely, is able to find ever new configurations of domination and submission in which to grow and thrive. Hence, if we eliminate specific forms of hierarchy without eliminating hierarchy in general, we may find that new hierarchies merely replace the ones abolished, while old hierarchies dig their heels in deeper. However, the general idea of non-hierarchy, while sufficient in its scope, remains insufficient in its differentiation and focus. The call to abolish 'hierarchy in general' must in turn be developed into a specific interpretation of social and ecological transformation. As it stands alone, the idea of non-hierarchy or cooperation remains too broad and ambiguous to have specific meaning. We are left wondering: What *forms* of non-hierarchy or cooperation are required? Unless we bring the idea of non-hierarchy into its specific fullness, we will be unable to translate it into a tangible social vision or practice. In the same way, without bringing the idea of boiling water into its specific fullness, we are left with an incoherent pile of necessary factors such as pots, water, and heating coils. We are left with little understanding of the relationship between the pot of water, the heating coil, and the synchronicity of time and place.

There exists a potentially complementary relationship, then, between that which is necessary and that which is sufficient *if and only if* all necessary conditions are consciously coordinated and integrated. Often, when people are overwhelmed by the complexity and urgency of social and ecological crises, they express frustration at the imperative to create a coordinated sufficient condition. They may reason, "Well, as long as we all do our own little necessary part, then eventually it will all form a sufficient whole." Such a response, while again understandable, fails to convey that if we each choose a necessary part, without consciously integrating those parts into a larger sufficient whole, we will keep the social project from realizing its full potentiality.

It is insufficient for one group to fight racism over here, while another group struggles against toxic dumping over there, while still another individual organizes a food coop some place else. This kind of 'piece work' is insufficient because it is non-holistic. When we see our activism as a series of single issues, we end up arranging the pot, the water, and the heating coil in different places at different times, failing to form a coherent vision of what we are striving for: a dazzling image of society boiled over, making room in the social stew for ever new revolutionary possibilities.

Once we have asserted the general idea of non-hierarchy as the integrated and coordinated sufficient condition for a free society, we may draw out the many necessary and specific forms of non-hierarchy needed to remake society. By differentiating the idea of non-hierarchy, we begin to educe a fully differentiated vision and plan for social and political reconstruction.

The Spheres Of Our Lives: Where Hierarchy Lives

In order to move toward a reconstructive vision, we need to comprehend the structure of the society we wish to transform. Just as the idea of non-hierarchy must be fully differentiated to understand the complex quality of institutional power, the idea of society must also be fully differentiated in order to convey the specific *locations* of institutional power.

When we think of society, we rarely think of the distinct spheres which give shape to our everyday lives. We usually refer to society as a monolithic structure as if we lived in a completely undifferentiated societal realm. Yet society is constituted of three distinct realms: the social, the political, and the State.

The social sphere is comprised of community and personal life. It is the sphere in which we create the everyday aspects of our existence as social beings. It is the realm of 'works and plays', the place in which we engage in production and distribution, fulfill community obligations, attend to practices of education, religion, as well as participate in a range of other social activities. While there is a public dimension to social activities such as work, school, and community life, there is also a private or personal component to social life as well. This is the space in which we reproduce the conditions of our most immediate physical and psychological needs and desires for food, love, sexuality, and nurturing. The personal dimension of the social sphere represents a specific quality of privacy predicated on an intimate knowledge of ourselves and of our closest relations.

In contrast, the political sphere is the space in which we assert ourselves publicly as managers of our own community affairs. It is the space in which we discuss, decide upon, and carry out the public policies which give form to social and political practices of our communities. The political sphere constitutes a specific quality of action which is distinct from the social sphere. Marked by a quality of public responsibility, the political sphere is the place in which we, as citizens of a town, village, or city, participate in shaping the policies which in turn inform our everyday lives.

Clearly, this description of the social and political spheres represents a brief sketch of what these spheres *ought to be*, rather than *what is* within our current society. Today, these spheres are dominated and degraded by the sphere of the State. The modern Republican state represents a hierarchical and centralized institution that both invades and appropriates activities that should be managed directly by citizens within the political sphere. The State coopts the power of citizens to directly determine and administer public policies regarding community activities such as production, technological practice, health, and education. To secure its own power, the State wields an often undetectable, yet constant, everyday threat of violence manifested through an army and police force.

The State has so thoroughly appropriated our understanding of 'government' that we are scarcely aware of our estrangement from truly autonomous *political* activity. Taking the State for granted as inevitable, we retreat into the social sphere looking for a site of both survival and resistance.

The Public Sphere: The Necessity Of Political Reconstruction

Yet in order to transform society, we cannot retreat into our social lives; we must address political questions as well. However, most social activists fail to sufficiently include the problem of reconstructing the political sphere within their activist vision. Instead, they often focus exclusively on the public and private dimensions of the *social* sphere.

The reasons for this are two-fold. First, the political sphere has been replaced by what Murray Bookchin refers to as "Statecraft": a system in which political power is placed in the hands of elected representatives (professional politicians) who make decisions regarding public policy on behalf of a voter 'constituency'. Disempowered by statecraft, and unaware of a political alternative, activists often turn away from questions of politics, turning instead to the social sphere where they feel they can at least exercise some control over their lives.[1]

Second, activists often neglect the political sphere because, estranged from their political identities, they identify primarily as consumers. The emergence of post-war consumer society gave rise to a generation of Americans who identified themselves through their consumption patterns. For instance, within the ecology movement, activists often identify more as consumers and technology 'users' than they do as political *citizens*. As a result, they tend to express resistance in the form of consumer activism by attempting to select, produce, or boycott particular commodities to establish congruence between their personal and political values. In this way, political power is reduced to 'buying power' as activists focus on questions of production and consumption rather than on trying to regain the political agency to determine what and how their community should produce.

For these two reasons—a politics reduced to statecraft and a political identity reduced to a consumer identity—activists tend to frame their opposition within social, rather than explicitly political terms. Within the social sphere, they feel empowered to make qualitative personal and social changes by improving the quality of their relationships with friends and family, improving schools and churches, or by creating economic alternatives such as coops or systems of community currency or barter. What is more, activists often unknowingly conflate social action with political action. Working to create social change within the domains of sexuality, spirituality, education, economics, and health care, they refer to this work as 'political', rather than social, as a way to emphasize the importance of the particular issue at hand or the necessity of changing public policy related to the issue.

For instance, members of such social organizations as Earth First! or Greenpeace are often referred to as 'political' organizations. Yet all members, from financial supporters to grassroots activists who participate in local and global campaigns, exist within a distinctly social, rather than political, relationship to one another. Again, political activity is that which takes place within the public sphere as citizens come together to discuss, debate, and determine the public policy that shapes their lives as members of a town, village, or city. Greenpeace, then, does not engage in *politics* in the literal sense. Instead, they wield crucial *social* contestation to state and corporate policy.

Social change is, indeed, crucial but without an actual transformation of political practice, we will never be in the position to actually determine the very economic, social, and ecological policies for which we are fighting. Instead, we will always be treated as children incapable of making our own decisions, forever appealing to the authority of parental representatives to do 'the right thing'. Temporary triumphs might be won; like little children who throw a tantrum to bend the will of their parents, we may beg our representatives to provide us with affordable housing or better environmental policy. However, the power relationship remains the same. The fact is, until citizens are able to make their own public policy regarding social issues, there will be no justice. We will be forever little children, tugging and whining at the hems of our parents' coats, begging them to make good decisions on our behalf.

Hence, our oppositional work is drained of its full potential as we linger along the periphery of the political realm, focusing mainly on social issues. In this way, we are weavers dreaming of beautiful tapestries, spinning and dyeing wool, envisioning clothes to be collectively woven and distributed, unaware that, without actually getting our hands on the equipment, our dreams will go unrealized. Direct democracy is the very process by which we make our dreams for a free society come into being. Without walking into the place where the cloth is woven, we will never be able to take those threads into our own hands to weave more cooperatively and more ethically. Instead, we will be left to wander about sheering, spinning, dyeing, and merely dreaming of beautiful shimmering cloth. Without walking into the public sphere, taking the power to make decisions into our own hands, we will be left to merely dream of freedom.

Illustrative Opposition: Illustrating The Political Implications Of The Social

Recognizing the necessity of political reconstruction leads us to look toward a process of political re-empowerment. Social ecology provides a thoughtful and comprehensive interpretation regarding how to engage in a political revolution by engaging in local municipal politics to initiate a broader move toward a confederation of directly democratic communities. Murray Bookchin's theory of *libertarian municipalism* proposes such a vision, offering a glimmer of hope for true democracy in a world where the political sphere has been hollowed out by the State.[2]

However, we confront a paradox when we consider the necessity of focusing on political reconstruction. While it is crucial to reconstruct an authentic political sphere, there will remain immediate social crises which also demand our attention. Clearly we cannot wait to address social issues such as homelessness, environmental racism, or violence against women until we have established a confederation of self-governing communities.

Illustrative opposition is way to focus upon a particular social issue *while* illustrating a broader political critique and reconstructive vision. In addition to demonstrating the *necessity* of a particular social issue, we may also illustrate the general *sufficient* condition required to fully address the particular issue at hand. For instance, early ecofeminist activists practiced a nascent form of illustrative opposition in the Women's Pentagon Action of the early 1980s. Beginning with an initial focus on the crisis of nuclear power, ecofeminists illustrated a wider social and political picture, drawing out broader issues of racism, capitalism, nationalism, militarism, male violence, and state power.[3]

Illustrative opposition must be specific enough to be meaningful, yet broad enough in order to deepen political consciousness. Had the Women's Pentagon Action presented too wide a focus, both participants and media would have been bombarded by the interconnecting issues of social and ecological injustice. However, had they focused too narrowly, say, on the ecological devastation of the earth by nuclear technology, they would have missed the opportunity to illustrate the widest implications of the nuclear crisis. The Women's Pentagon Action was successful in broadening an understanding of the necessary conditions for creating a nuclear-free society. Through theatrical demonstrations and written media, these early ecofeminists helped others to explore a range of necessary conditions pertaining to the spheres of the social and the State by demanding an end to racist and masculinist state practices in regards to nuclear energy and militarism, and by confronting capitalist production of nuclear technologies.

However, while the Women's Pentagon Action presented an extensive critique of the spheres of the social and the State, like most movements of the New Left, they failed to extend their critique to the political sphere. By linking a critique of social and state institutions to a demand for direct democratic control over social *and* political life *in general,* the Women's Pentagon Action would have presented a sufficient condition for a nuclear free society.

In this way, illustrative opposition is a practice of holistic picture-making in which one brush stroke serves as an invocation to bring an entire picture to fullness. The idea of holism, inherent within the idea of illustrative opposition, conveys that a whole is not just the sum of its parts. For instance, in the case of the pot of boiling water, the whole, or the boiling pot of water, is not reducible to the pot, to the water, or to the heating element. Accordingly, it is insufficient to simply throw the necessary parts together in a room, expecting to bring water to a boil. As we have seen, it is the specific and irreducible *relationship* between the parts that gives the whole its particular form and function. It is the specific and irreducible relationship between individual forms or parts of oppression, which gives the whole oppressive system its form and function as well. Hence, the goal of illustrative opposition is to focus on one part of a larger system of oppression to depict a whole which is appreciated in its interconnected complexity.

THREE MOMENTS OF ILLUSTRATIVE OPPOSITION

Illustrative opposition unfolds in three phases. In the first *critical moment*, we recognize a particular form of social or political injustice, responding in turn with critique. In this moment, we may sort through the separate strands which compose the central cord of a particular form of injustice. We may analyze how this form of injustice surfaces and is perpetuated within realms of the social, the political, and the State. In the critical moment, we ask ourselves what makes this particular form of injustice unique or particular, asking: How is this form of injustice different from other injustices; why has it become a crucial issue at this point in time; or what makes it historically unique?

In the critical moment, we look at the historical development of the particular issue, examining in turn, the lesser known radical history which surrounds the form of injustice. Hence we would ask: Were there attempts in the past to resist this form of injustice; what made these attempts successful or unsuccessful; what is to be learned from both the history of how this injustice came to be, and the history of what almost was, or would have been?

In the second *reconstructive moment,* we begin to draw out the wider reconstructive potential nascent within the struggle against a particular form of injustice. We begin by examining the implications of engendering wider conditions of justice surrounding the issue within the realms of the social, political and the State, examining in turn, the ecological implications of the particular injustice at hand for each sphere. Here, we explore how to transform each sphere of society sufficiently in order to thoroughly transcend the particular form of injustice. Ultimately, the reconstructive moment serves as an opportunity to draw out the social and political conditions that are necessary to sufficiently oppose and transcend the particular form of injustice.

Finally, the third moment constitutes the *illustrative moment.* Here, we begin to elaborate ways to articulate and demonstrate the many insights we glean as we move through the previous moments. There are many forms through which we may express these comprehensive insights: We may print pamphlets which are critical, historical, and reconstructive in nature; develop a performance piece that integrates our insights and conclusions; take direct action, creating banners with slogans that point to salient threads of our overall analysis or vision; articulate our analysis on alternative and mainstream media such as pirate radio or the Internet; or create teach-ins and ongoing lecture-discussion series within our communities.

Our 'illustrations' must be utopian and visibly socio-erotic. For our goal is not only to inform, but to inspire ourselves and others to take direct action. As previously discussed, we need to restore to the erotic a distinctly social meaning, articulating the different 'moments' or aspects of social desire, cultivating a language to describe our yearnings for community and association, creativity and meaning, self and community development, and social and political opposition. Such yearnings stand in sharp contrast to the vernacular understandings of desire that are framed in terms of individualized accumulation of status, power, or pleasure. To understand the socio-erotic is to locate moments of individual desire within a distinctly social and political context, appreciating the potential of our individual desires to be accountable to, and enhancing of, a greater social collectivity.

Our illustrations must speak to our socio-erotic desires. Within the bland culture of global capitalism, people crave authentic integral sensual stimulation. The appeal of theater groups such as Bread and Puppet attests to the sensual power of creative media. The display of towering and colorful puppets parading down barren city streets during demonstrations summons up the sensual awe and desire for our own creativity in a world of commodified alienation, allowing us in turn to remember our own creative potential. We need to appeal to as many media as possible to illustrate our analysis and vision, utilizing art, theater, dance, electronic media, print media, speak-outs, and street demonstrations, illustrating the sensual presence and resistance of our physical bodies as well. In this way, illustrative opposition must be sensual: it should constitute the ultimate body politic in which we literally throw our bodies into social contestation, taking illustrative and expressive direct action. However, such actions must not only 'show' but they must also 'tell' a narrative, moving from the particular to the general or from the personal to the social and political. People join social movements for a variety of reasons. In addition to wishing to transform the world, activists often yearn to transform themselves. They come to movements out of associative desire: out of the desire to find friendship, lovership, community, and meaning. Seeking a sense of connection and purpose, people are drawn to particular social movements because the people within the movements embody the intelligence, passion, and communality they wish to develop within themselves. Hence, our illustrations must convey both the values of the world we want to create as well as the values of the people we want to draw into our movements. While our work must be collective and non-hierarchical, our forms of contestation must put forth a display of communality as well. We must clearly articulate the ways in which others may join our struggle, continually illustrating points of entry into our social movements.

Further, we must address our creative or differentiative desire as we illustrate our opposition. In this age of incoherence, we each have an underlying desire to differentiate, or to 'make sense' of the chaos which surrounds us. As we are overwhelmed by social, political, and ecological crises, we yearn for illustrations that render our world more legible and intelligible. Our illustrations must draw what is coherent and clear out of what is confusing and opaque. The goal of illustrative opposition, then, is to help others to literally 'sort out' the different spheres of social and political injustice, bringing others to a state of increased confidence and desire for ever greater levels of understanding. Hence, our illustrations must be educational as well as sensual and associative; they must represent ongoing teach-ins in which we assist ourselves and each other to recover lost radical history and a rational and coherent analysis of injustice.

In turn, we must consider our developmental desire as we create new expressions of social opposition. Developmental desire represents the yearning of the self to become more of itself, to uncover ever wider horizons of competence, joy, and community. Our illustrations must represent opportunities for self-development in general that offer more opportunities for participation than spectacle-gazing. Through social contestation, we may develop abilities for public speaking, writing, teaching, and art-making; we may become lecturers, poets, and painters, speaking at coffee houses, concerts, universities, street corners, community health centers, libraries, cable stations, and city halls, creating a counter-spectacle of coherent disruption.

Finally, our illustrations must inspire oppositional desire. Far from the individualistic and acquisitive desires that constitute our everyday lives under global capitalism, we need to publicly articulate and express a new vision of desire: a social desire, a desire informed. Engendering a new oppositional desire is a potent antidote to an age of authority-induced passivity. Corporate CEO's and state agents dismiss our rants about 'desire'—as long as we keep our desire bound within the social sphere. Once we draw out the political implications of desire, informing our desire with a rational demand for direct participation in determining the conditions of our everyday lives, *then* we will see real opposition and fertile conflict.

Illustratively Opposing Biological Patenting

We may begin to think through a potential illustrative opposition by addressing a particular form of social injustice: the patenting of human and biological life. Beginning with an ecological problem that touches upon realms of the social and the State, we may transform this problem into a point of departure, a seed out of which we may draw a wider analytics of revolutionary political reconstruction. We may begin by taking a brief look at the issue at hand, then explore a series of questions that may lay the ground for a deeper understanding of the sufficient condition for a 'patent free' society.

Problem Background: What Are Biological Patents?

Within the world of biotechnology, a new vocabulary emerges that equates the genetic modification of cells to an act of 'creation'.[4] Just as Columbus 'discovered' and thus 'claimed' the New World of North America, a continent that had been home to civilizations of native people for thousands of years, biotechnologists are 'discovering', recombining, and laying claim to the cell-lines of plants, animals, and even human beings whose DNA might prove useful to such industries as agriculture, pharmaceuticals, or reproductive medicine.

The question of legal patents of cellular materials is one of the most controversial issues surrounding biotechnology. Historically, a patent gave exclusive rights to an inventor to exploit a product, process, or a particular use of a product for a limited time, usually ranging between 17-24 years. In order to obtain a patent, the product or process had to be *invented*. The precedent for patenting was established at the International Convention for the Protection of Industrial Property held in Paris in 1883, the first international agreement on intellectual property rights. By 1930, the Plant Patent Act permitted the granting of patents for plants reproduced by cutting or grafting to produce plant hybrids in the United States. Toward the end of the 1970s, as practices of genetic engineering through recombinant DNA became increasingly successful (and thus potentially commercially viable), a quiet war began to emerge between private corporations, patenting courts, and the Supreme Court regarding the right of individuals to patent a wider variety of life forms.

Beginning in 1971, the General Electric (GE) company embarked on the crusade to obtain the first patent for a non-plant life form. In 1970, GE engineer Ananda Mohan Chakrabarty developed a specialized bacterium that promised to break down or 'eat' oil from tankers spills. Over a period of ten years, GE and the Court of Customs and Patents Appealed (CCPA) waged a relentless campaign of litigation against the U.S. Patent and Trademark Office (PTO) and the Supreme Court to patent this oil-eating bacteria. Once patented, GE knew, the bacteria could set a precedent for future patenting of other life forms to be appropriated by biotechnology corporations.[5] In 1980, GE's oil-eating bacteria won its case as the Supreme Court granted Chakrabarty his patent. In this gesture, the Supreme Court determined life itself patentable, stating that "the relevant distinction was not between living and inanimate things" but whether living products could be seen as "human made inventions."[6]

As predicted, GE's Chakrabarty case opened the floodgates for the budding biotechnology industry. That same year, emerging biotechnology industries such as Genetech and Cetus took Wall Street by storm, setting records for the fastest price per share increase ever. The burgeoning biotechnology industry inspired other corporations and scientists to patent not only microorganisms, but plant, animal, and even human life forms as well.[7]

Presented as a solution to urgent problems of disease or world hunger, biotechnological inventions also 'solve' capitalists' 'need' for profit and growth.

The development of the new biotechnology is controlled primarily within capitalist structures such as transnational enterprises, universities funded by corporations, and small 'start-up' corporate firms. Already, biotechnology has been applied in primary industries of agriculture, forestry, and mining; in secondary industries of chemicals, drugs, and food; and finally in tertiary industries of health care, education, research and advisory services.[8]

Addressing The Question Of Biological Patents

When a group sets out to address a problem such as intellectual property rights, or biological patenting, the group faces a crisis so complex and overwhelming that to merely address the particular problem at hand seems insurmountable. For instance, indigenous communities in the Amazon engaged in fighting the patenting of local medicinal plants by transnational biotechnology corporations, are already often so involved in other struggles for survival that contestation often focuses on protecting indigenous communities from the specific harm of biological enclosure.[9]

Accordingly, questions of biotechnology are often cast within the terms of the offending party itself, framed in *social* terms of economics and production (as groups resist particular corporate practices), in terms of state power (as groups address national and international patenting policies); and in the social-statist terms of international trade (as groups deal with international trade agreements facilitated by the World Trade Organization (WTO). Yet for contestation to such practices as biological patenting to be rendered sufficient, they must be understood not solely in the terms of *freedom from* specific injustices within the realms of the social and the State, but in terms of *freedom to* create a socially and politically free society in general.

How can we reason from a particular crisis such as the patenting of living organisms to reach a general analysis of social and political transformation? How can we reason from the dystopic crisis of life patenting to a vision of a world that is not only patent free, but is free of all forms of hierarchy in general? What follows offers a brief outline, a set of illustrative and oppositional questions that allow us to begin to reason from the particular to the general, from the social to the political, and ultimately, from the ecological to the revolutionary.

I. The Critical Moment

In the critical moment, we begin explore the social and statist dimensions of life patenting. We initially ask: How does the patenting of biological life inform the social sphere, both public and private? Beginning by looking at the *private* dimension of the *social sphere*, we might ask: If the most basic and organic unit of private life lies within the body itself, then we may explore how the body's autonomy and privacy are degraded by patents that impose new capitalist relations within the very germ plasm of life. As we attempt to critique the private dimensions of this crisis, we need to look for historical novelties, asking: What makes this form of injustice distinctive and new? By addressing such questions, we examine the particular implications of patenting for private life in general, exploring novel ways in which patenting disrupts bodily integrity, reducing cell-lines to marketable materials to be owned and hoarded by corporations.

Next, we would critique life-patenting in relation to the *public* dimension of the *social sphere*. Here, we would explore such issues as capitalist production, consumption, and public education as they relate to biotechnology. We may point to moments of commodification and ownership of life forms as well as corporations' search for ever new colonies (biological as well as social) for never-ending expansion. As we recognize the particular urgency of this crisis, we may point to what makes this particular crisis distinctive, asking: What makes biotechnology different from, and potentially more harmful than, other forms of commodified scientific practice? Or, what makes life patenting different from other forms of colonialism? Or, how does the imperialistic devaluation of local indigenous knowledge and life itself 'legitimize' the patenting of species used in indigenous agricultural and medicinal practices?

As we critique the implications of patenting for the social sphere, we may explore the novel impacts of such practice on institutions of public education. Here we may explore how patenting practices inform research agendas and funding priorities within microbiology departments in universities throughout the United States and much of Europe. In particular, we may begin to examine the increasingly intimate relationship between publicly funded research and private industry.[10] This relationship is changing dramatically as public universities grow increasingly dependent on private industry for funding, and as biotechnology industries become attractive and socially accepted research arenas for scientists. We must explore the implication of scientific practice within a context in which increasingly, scientists conduct research out of personal economic interest, rather than out of the 'love' of 'pure' science.

When we engage in the critical moment, we may also show moments of resistance which show the limits of hegemony itself. For instance, we would explore how in India, farmers have engaged for years in an ongoing struggle against World Trade Organization (WTO) proposals on agriculture and intellectual property rights which would allow transnational companies monopolize the production and distribution of seeds and other aspects of Third World agriculture. We might explore an earlier struggle, in October of 1995, in which a half-million Indian farmers from Karnataka took part in a day-long procession and rally in the South Indian city of Bangalore, constituting the largest display of public opinion anywhere in the world either for or against the round of Geneva trade talks surrounding the WTO. At this event, Karnataka farmers established an international research center in order to help develop community seed banks and to protect the intellectual rights of their communities.[11] It is vital to uncover the rich moments of resistance such as these that are scattered across the globe. We need to continually shed light on movements of social contestation that bubble up amidst even the most oppressive conditions. In this way, our critique is informed not only by urgency, but by vital inspiration.

Further, we may critique the *sphere of the State* surrounding patenting. Here, we examine novel articulations between the State and the social sphere, exploring how state institutions including the National Institute of Health and the Department of Energy fund social institutions such as corporations and universities to collect, taxonomize, and warehouse genetic information through such projects as the Human Genome Project (a three billion dollar program that is currently 'mapping' the entire human genome).[12]

Finally, we may pose a series of critical questions relating to the *political sphere* concerning the lack of popular awareness and participation in determining public policy surrounding life patenting. Here, we critique the lack of scientific literacy among citizens, the lack of public forums for popular education, discussion, and debate about current scientific practices. Here, it is crucial to draw out the general crisis surrounding non-democracy from the particular crisis of biological patenting.

In the critical moment, we may explore the historical context of life-patenting by examining the radical history of resistance movements related to the topic more generally. We might begin by looking at the historical relationship between public and private institutions of science, medicine, education, and capital, examining the theme of colonization and privatization. Particularly, we would examine the historical context surrounding intellectual property rights, looking at the roles institutions have played in developing such practices over the century. We would also analyze the broader history of colonialism, capitalism, and patriarchy that frames such issues as seed cultivation and ownership in Third World situations. We would look at the legacy of the nation-state in the colonial and neo-colonial eras, examining the breakdown of local indigenous self determination of social and ecological policy.

In turn, we would explore the history of resistance to life-patenting. We would explore movements throughout the Third World that have continued to resist capitalist enclosure since the first phase of colonialism. In order to reveal this radical history, we would need to uncover the historical continuities between resistance to current life patenting practices and to previous expressions of colonial enclosure. In this spirit, we would generalize upon the particular meaning of life-patenting, tracing the emergence of anti-imperial movements which contested injustices such as slavery and land enclosure.

II. THE RECONSTRUCTIVE MOMENT

In the reconstructive moment, we begin to consider the liberatory possibilities presented by addressing the particular form of injustice at hand. In the reconstructive moment we treat the three spheres of society differently: while we look to *transform* the social and political spheres, we examine avenues for *transcending* the sphere of the State.

Beginning again by looking at the implications of biological patenting for the *social sphere*, we may explore the reconstructive possibilities of revaluing the private dimension of the body. In the reconstructive moment, we begin to highlight the continuities between particular and general forms of injustice. For example, while life-patenting introduces particular novel legal, cultural, and corporate practices related to private 'embodied' dimensions of life, it also builds upon a more general history of privatizing human and other life forms. It is consistent with a capitalist 'tradition' which enslaved African Americans in the American South, bound women legally to their husbands, and continues this tradition by trafficking women and babies in sex industries and black-market adoptions, in addition to commodifying land, plants, animals, and other organisms.

Here we understand that the sufficient condition for reclaiming the body and 'life' itself, is to abolish the practice of patenting in *all* spheres of society. A truly free society entails that no body, person, or organism can be reduced to private property, no human can be rendered subject, either in part or in entirety, to another person or institution.

As we continue to think through the social sphere, we may consider what it would take to create social and political conditions which render all forms of private property (bodily or otherwise) unacceptable. Exploring the role that medical, pharmaceutical, agribusiness, and chemical companies play in determining research and regulation of genetically modified organisms, we would look to remake the social sphere along post-capitalist lines.

What is most crucial in the reconstructive moment, then, is to draw out the most utopian and sufficient conditions of freedom which surround a particular issue. For instance, while it is necessary to eliminate patents of biological life, we must illustrate how merely abolishing such patents represents an insufficient condition to engender a truly free society in general. We would point to the widest conditions of freedom that can be drawn out from the idea of a patent-free social sphere. We would begin to articulate the need for a sphere of education, technology, and economics that is based not on commodification, but upon social cooperation.

As we consider transcending the *State* we may begin to draw connections between the particular form of injustice in question and the lack of direct democracy throughout society as a whole. It is vital to articulate specific ways in which current state governments inhibit citizens from participating directly in determining the policies that affect their lives. In turn, we must also show how the lack of confederal forums deprives us of informing the unfolding of events outside our own municipalities and throughout the world.

In thinking through the issue of life-patenting, we recognize that disruptions caused by such practices are not exclusively local in nature. Within the age of global capital, we see that there exist few uniquely local problems as currents of capital and state power flow throughout towns, cities, states, and countries the world over. Although corporate, governmental, and regulatory institutions that control the collection and storage of genetic materials operate within specific localities, these institutions function within an international system of trade, production, regulation, and policy making which is *transnational* in character.[13]

In the reconstructive moment, we would begin to explore how to transcend the State by creating a new politics in which citizens have direct control over technological practices such as biotechnology. We may illustrate how, by replacing the State with a confederation of directly democratic municipalities, citizens would empower themselves to discuss and decide scientific matters that affect not only organisms and people locally, but globally as well. In the reconstructive moment, then, the criticism and analysis of a particular form of hierarchy opens the way to elaborate the broadest understanding of non-hierarchy possible.

III. The Illustrative Moment

The illustrative moment represents an opportunity to inspire others to demand the sufficient social and political conditions for a free and ecological society. It is the forum in which we inspire others to move beyond the scope of a particular crisis, to demand self-determination within a broader political context. It is the moment to create oppositional forums in which we may ask: What does life patenting have to do with democracy? Or, what does abolishing patenting have to do with creating a utopian society?

Illustrative opposition should compel ecological activists to reach for new connections between social and ecological issues and their authentically *political* implications. Each moment of illustrative opposition to state practices for instance, should point to the wider demand for authentic direct democracy. Illustrative opposition allows us to highlight a particular moment in which we have no direct political control, raising awareness of our lack of policy-making control *in general*. The illustrative moment explains by asking questions.

Through our actions and our propaganda, we ask: how did it come to be that we control so little regarding this particular issue and regarding our lives in the broadest sense?

There are many ways to illustrate the need for direct democracy. As discussed earlier, we can popularize the demand for political power using a variety of media ranging from radio, pamphlets, and teach-ins to guerrilla theater, bill board alteration, and murals. There is no 'recipe' for making the connection between ecological and revolutionary *political* issues, as each activist group brings their own talents and sensibility to the project of opposition.

I am a member of a small media collective in Western Massachussetts that engaged in illustrative opposition regarding issues of biological patenting and agricultural biotechnology. Last year, the group saw the need to raise public awareness regarding the introduction of genetically engineered organisms into the food supply that has begun in recent years. In addition to being concerned by insufficient research on the potentially allergenic and toxic effects of ingesting genetically engineered foods, we were troubled by the lack of research regarding environmental risks that surface as plants spread their genetically engineered traits to other neighboring organisms (through cross-pollination or ingestion).

But we were not solely concerned with environmental and health risks associated with genetically engineered crops. The group also wanted to address issues of economic and cultural self-determination surrounding the issue. We wanted to educate ourselves and the public regarding how local farmers throughout the world are economically and culturally threatened as multi-national agro-chemical companies gradually monopolize the seed industry worldwide.

We also had another primary concern. Our group wanted to illustrate the link between the social and ecological problems presented by genetically engineered crops and the need for political transformation. We wished to demonstrate how both corporations and the State, rather than citizens, determine economic, ecological, and political policy related to agricultural biotechnology. As a media collective composed of writers, actors, and artists, we decided to create a series of theatrical events as a way to illustrate our opposition to biotechnology.

At a demonstration that protested Monsanto (a U.S. based multi-national agro-chemical company heavily invested in biotechnology) corporate offenses, our group performed a theater piece in which a two-headed monster (wearing name-tags that read "the State" and "Capitalism") delivered an oratory regarding its autocratic decision to find new avenues for capitalist expansion through biological patenting and genetic engineering. Surrounding the monster, floated a sea of zombie-like people (wearing signs that read "consumer") who stared blankly and passively at the monster as he announced his plan. Over the course of our skit, the consumers first strolled about passively, then attempted to fight the monster, and finally ended up gathering together to discuss what to do next. Through this process, the consumers realized that by gathering, discussing, and making decisions, they had actually formed a town meeting of sorts, and they realized that what they really wanted was to reclaim their political power. One by one, the consumers flipped over their signs to reveal the word "citizen" written on the other side.[14]

At the end of the piece, the actors sat in a circle and invited the audience to join them in an impromptu town meeting to discuss plans for continuing the struggle for direct democratic control over technology and over life in general. What actually ended up occurring, though, was a more concrete, yet highly democratic discussion of plans for the anti-GMO movement itself.

We then did a series of "supermarket inspections" in which we dressed in white bio-hazard suits to go 'shopping' at our local supermarkets. We strolled down the supermarket aisles, 'inspecting' the produce with a variety of bogus scientific instruments, dropping flyers into people's shopping carts and into produce and dairy displays. In addition, each 'inspector' (unable to speak through a gas-mask) had a plain-clothed 'assistant' who would strike up conversations about biotechnology and democracy with other shoppers whose responses ranged from amusement and interest, to suspicion and annoyance. During each action, we had between five to fifteen minutes before we were asked (or aggressively forced by security guards) to leave the store.

In our flyers, we explained that we were a renegade group that had defected from the Food and Drug Administration after deciding that we desired direct political power—in addition to 'safe food'. Discussing the economic and cultural issues associated with genetically engineered foods, the flyer also talked about the connection between direct democracy and technology, attempting to raise the level of public discussion from questions of environmental and health risk to issues of political power.

For our next action, we plan to set up a "patent office" on a busy street in our town where we will hand out patent applications to passersby, offering them the chance to patent their own cell-lines. Through satire, we plan to educate members of our community about biological patenting, both human and non-human, explaining the relationship between issues of bodily integrity, social issues such as capital-driven biotechnology, issues of state monopoly over policy making, and political issues such as the need for direct democratic control over technology and over our lives in general.

Through these small actions, we are trying to widen the discussion surrounding biotechnology by talking about questions of political power in addition to issues of environmental and health risk related to genetically modified foods. It is our hope that people may begin to see themselves as more than *consumers* seeking the power to buy safe food. We want to encourage people to see themselves as *citizens* who desire the *political* power to create a humane and ecological society.

In turn, we are hoping to move discussions surrounding biotechnology beyond romantic yearnings for a golden age untainted by 'technology'. In our actions, the idea of 'nature' is taken from the realm of abstraction and is brought down to the realm of the everyday. The 'nature' we invoke is our bodies walking down a city street and it is the food we buy in the supermarket. In turn, we show that the cause of ecological injustice is not abstractions such as 'civilization' or 'industrial society'—but rather, a set of social relationships called the State and capitalism that appropriate our power to create cooperative relationships within society and with the rest of the natural world.

Our group has just begun to think through the process of illustrative opposition. As a collective of actors and writers, we have chosen to express our opposition in the form of theater and written text. But as I mentioned earlier, dissent has a variety of forms. By giving a brief sketch of some of our first actions, I have tried to depict a 'work in process' that aims only to stimulate conversation, critique, and perhaps action as well. As our group continues to explore the relationship between direct democracy and technology, our actions will hopefully embody an increasingly elaborate understanding of the necessary and sufficient conditions for creating a free and ecological society.

As our group knows, revolution cannot be generated from a series of individual protests against social and ecological injustices. It requires that we articulate not only what we do not want, but what we desire as well. The demand for *substantive freedom*, or the demand for the very substance of what freedom means, stands in contrast to the demand for *negative freedom*, which while necessary, represents an incomplete demand to negate injustice. We must be able to articulate a substantive vision of the society we desire, illustrating through our activism, the social and political freedoms for which we yearn. We must illustrate a substantive demand for the *freedom to* create a society based on a confederated direct democracy, a municipalized economy, and on a new social and ecological sensibility based on values of cooperation and mutual-aid.

Through illustrative opposition, we are neither locked into single-issue activism, nor locked into the stagnation of 'waiting' for a local or national political movement sufficiently comprehensive to address the widest range of revolutionary desires. To be sure, we cannot sit back and watch urgent crises pass before our eyes. Instead, we may address the *necessity* of a single issue, presenting a wider *sufficient* condition for a free society in the process. Thinking through each particular moment of unfreedom opens the way to consider the widest vista of freedom imaginable.

It is vital that we begin to think along coherent revolutionary lines. In this age of incoherence, our thinking about social and political change often tends to be scattered and fragmented. The spectacle of the nightly news does not assist us in understanding the crucial link between real political power and the struggle for social and ecological justice. Instead, we are expected to sit back and watch the parade of incoherent events presented to us as disparate and unrelated as the commercials that flicker by every four to seven minutes.

To create coherence in the age of incoherence is a highly oppositional act. By clearly conveying the 'logic' that underlies this irrational world, we actually lessen the overwhelming burden of social disorientation. To see how one crisis emerges from the other—to think rationally—opens the way to understand how one phase of reconstruction may emerge from the other allowing us to gradually transform society as a whole.

A crucial component of any illustrative opposition is a process of education in which we recover a sense of theoretical and historical integrity. In this spirit, we may create study-groups and centers for radical education, forums in which we may think through the moments of illustrative opposition, educating ourselves in revolutionary history, awakening ourselves to the possibilities for social and political reconstruction.

Illustrative opposition, then, is not merely an instrumental means-ends approach to social or political activism. Rather, it represents a comprehensive and utopian analytics made visible. The illustrations that we paint represent valuable ends in themselves; they represent an ongoing challenge to the institutions that oppress us, a challenge that shows the world that opposition is alive, well, and will not go away. Our illustrative actions must curb the steady tide of social and political injustice that gathers strength daily. As we begin to popularize the demand for direct political power over our everyday lives, the horizon of social and ecological justice no longer recedes into the distance, but rather, calls out to us, yearning passionately for its own actualization.

Notes

1. For a wider discussion of the distinction between statecraft and authentic political practice, see *Urbanization Without Cities: The Rise and Decline of Citizenship,* (Montreal: Black Rose Books, 1992), pp. 123-175.

2 . Libertarian municipalism represents the political vision of social ecology, a body of philosophical and political theory developed by Murray Bookchin. Beginning in the 1950s, Bookchin, a libertarian socialist himself, began to create a synthesis of Marxist and left libertarian thought, addressing problems raised by gender oppression, ecology, and community as well as addressing the new developments of capitalism. He then went on to formalize a coherent theory of the social origins and solutions to ecological problems, establishing himself as perhaps the most prominent 'leftist voice' in the ecology movement, a role to which he is still fiercely committed today. His theory of libertarian municipalism represents an interpretation of how to gradually transform the current nation-state into a confederation of direct democratic municipalities, drawing upon the libertarian dimensions within the French and American revolutionary traditions. For a cogent and compelling introduction to the idea of libertarian municipalism, read Janet Biehl, *The Politics of Social Ecology* (Montreal: Black Rose Books, 1998).

3. Ynestra King, a primary organizer of the Women's Pentagon Action, gives an excellent description of the kind of illustrative and ecological thinking which surrounded the event. See "If I Can't Dance in Your Revolution, I'm Not Coming," Adrienne Harris and Ynestra King, eds., *Rocking the Ship of State* (Boulder: Westview Press, 1989), pp. 281-298.

4. According to Vandana Shiva, "Biotechnology, as the handmaiden of capital in the post-industrial era, makes it possible to colonize and control that which is autonomous, free and self-generative. Through reductionism science, capital goes where it has never been before." For an excellent discussion of biological and cultural generativity, see Vandana Shiva, "The Seed and the Earth: Biotechnology and the Colonisation of Regeneration," in Vandana Shiva, ed., *Close to Home: Women Reconnect Ecology, Health, and Development Worldwide* (Philadelphia: New Society Publishers, 1994).

5. Pat Spallone. "The Gene Revolution," *Generation Games* (Philadelphia: Temple University Press, 1992), p. 120.

6. Andrew Kimbrell. "The Patenting of Life," *The Human Body Shop* (San Francisco: Harper, 1993), p. 195.

7. Indeed, the patenting of human cell-lines has led to some dramatic legal crises. In 1984, scientists at the University of California licensed a cell line taken from the spleen of leukemia patient John Moore to the Genetics Institute who, in turn sold the rights to a Swiss pharmaceutical company, Sandoz. One estimate places the long-term commercial use of Moore's genetic material, known as the "Mo Cell line" (patent #4,438,032) at about one billion dollars. In addition, Moore, whose permission had not been sought for the taking of his cells, demanded the return of his spleen cells before the California Supreme Court. In response, the court determined that Moore had no direct claim on his spleen cells but that he did have the right to sue doctors for not advising him of his rights. See Beth Burrows, "Message in the Junk: Commodification and Response." Paper presented at *New Currents in Ecological Activism Colloquium.* Institute for Social Ecology. Plainfield, VT. 1 July 1995.

8. Vandana Shiva, *Biotechnology and the Environment* (Pulau Pinang, Malaysia: Third World Network, 1993), p. 2.

9. For a wonderful discussion of the relationship between indigenous knowledge and intellectual property, see Vandana Shiva, *Biopiracy: The Plunder of Nature and Knowledge* (Boston: South End Press, 1997).

10. Paul Rabinow provides an ethnographic account of the relationship between private industry and genetic research in *Making PCR: A Story of Biotechnology* (Chicago: University of Chicago Press, 1996).

11. Martin Khor. "500,000 Indian Farmers Rally against GATT and Patenting of Seeds," *Resurgence,* Jan. 1993., p. 20.

12. For a particularly insightful discussion of the Human Genome Project, see R.C. Lewontin, "The Dream of the Human Genome," in *Cultures on the Brink: Ideologies of Technology,* Gretchen Bender and Timothy Druckrey, eds. (Seattle: Bay Press, 1995), pp. 107-129.

13. See Vandan Shiva, *Biopiracy.*

14. My thanks to Bob Spivey for developing what was truly, a wonderful script.

On an Ecology of Everyday Life

While ecological restoration is necessary, it alone is insufficient for reclaiming a desirable quality of social life. Ecology must evaluate the social, political, cultural—as well as the biological—dimensions of life, demanding the power for citizens to be able to determine the nature of their relationships with each other and with the rest of the natural world. An ecology of everyday life is a social ecology that translates the desire for "nature" into a politicized desire for direct democratic control through which citizens may create a society that is whole, humane, and meaningful.

We must cease to portray "nature" as a distant, pure, abstract thing removed from the everyday lives of people living in urban and degraded rural environments. It is time for "nature" to be brought down to earth, to become the very stuff of our lives: the crowded street in our neighbourhood, the water with which we wash our clothes, both sky scraper and smoke-stack, as well as the plants, animals, and other creatures with whom we share this planet.

To fulfil its revolutionary potential, ecology must become the desire to infuse the objects, relationships, and practices of everyday life with the same quality of integrity, beauty, and meaning that people in industrial capitalist contexts commonly reserve for "nature." It means recasting many of the values often associated with nature within social terms, seizing the power to create new institutions that encourage, rather than obstruct, the expression of a rational social desire for a cooperative, healthful, and creative society. The idea of nature can no longer be the "country home" of our desires, that place we run to in our dreams, longing for escape from the pain and confusion of life in the era of global capital. We must relocate the idea of nature within society itself, transforming society into a ground in which we may build, collectively, a new practice of both nature and community.

The call for an ecology of everyday life speaks not just to our immediate physical needs for survival. In addition, it arouses the desire for a world forged by social desire in all of its forms: a life redolent with personal creativity and a quality of community life based on humane and ecological practices. Ecology provides a lens through which we may take a long and often excruciating look at our own lives, a chance to evaluate the quality of our relationships, both local and global. And if we are not heartened by what we see, we realize that we have an enormous challenge before us. For once we appreciate the interconnectedness of life, we understand that we cannot simply work to save ourselves or a certain species of plant or animal—we realize that we must transform society as a whole.

The demand for an ecological society cannot be reduced to an individual or personal quest for a better quality of life. As I have tried to illustrate, an ecology of everyday life entails instead a rational social desire to establish a quality of life for all people, a desire that ultimately requires a dramatic restructuring of political, social, and economic institutions. It asks that we transform our love for nature into an activist politics that strives to bring to society the best of what we long for when we talk about "nature."

This requires that privileged people reconsider attempts to simplify their life styles, to, in addition, grapple with what I call "the complexity of complicity": a recognition that, despite the attempts of privileged people to extricate themselves from systems of injustice through personal life-style choices, because of the pervasiveness of overlapping systems of power, they will always remain embedded and thus complicit within such institutions as global capitalism, the State, racism, and sexism.

But instead of despising themselves for this privilege, or trying to assuage their guilt by individually trying to lead simple lives, privileged peoples might instead begin to redefine their guilt as "ineffective privilege." They may identify their privilege—whether it be based on physical ability, education, economic status, ethnicity, gender, sexual orientation, or nationality—and they may transform this privilege into a potent substance to be used for social and political reconstruction. Guilt associated with the privilege of money, race, and education, for example, may be transformed into time, economic resources, and information useful to political struggles. Privilege within complex systems of hierarchy can be morphed from paralyzing guilt into an active process of thinking rationally and compassionately about how to utilize particular resources to dismantle systems of power.

Recognizing the complexity of complicity means accepting that there are no simple or romantic escapes from the challenges that stand before us. We realize that instead of seeking comfort within a people-less wilderness, we must confront and rebuild social and political institutions—a task that entails a long-term struggle that is far from romantic. It requires that we embark upon the often arduous struggle of working with others to create ethical and rational political organizations and movements. An ecology of everyday life transforms ecology from a lofty romantic venture into an ongoing labour of love. Ecology is as much about the drudgery of licking envelopes for a mass-mailing and fighting to save an urban community center in the Lower East Side of Manhattan as it is about saving a forest.

Once we let go of romantic conceptions of desire, we are free to explore a social desire that rounds out our humanity, enticing us to become ever more sensual, cooperative, creative, developmental, and oppositional. We may recast our lives in social terms, recognizing desire as an anticipation of the pleasure that comes from enhancing the satisfaction and efficacy of both ourselves and others. Here, ecology becomes the light by which we scrutinize our everyday lives; it is the voice through which we demand the power to bring forth a world in which we may live the boldest and most social expressions of our humanity.

An ecology of everyday life entails rethinking our understanding of nature as well. Removing the idea of nature from its pristine and static display case, we may see nature for what it is: a dazzling and dynamic evolutionary process that continues to unfurl about us and within us. Once we are able to locate ourselves within this evolution, we can begin to measure our everyday lives as they are against what they could be if only we were free to actualize our potential for such evolutionary coups as cooperation, creativity, and development. Suddenly, the dull office job, the lonely neighbourhood, the poverty, or even the unsatisfying privilege—all take on new meaning. Rather than constituting a personal failure or a lack of will, our withered communities and lives reflect an anti-social and hierarchical trend that has spread through humanity like an industrial fire. By recognizing our minds, our hands, our bones, and our hearts as part of natural evolution—as an evolutionary inheritance—we become outraged by this fire, breathing it into our lungs, transforming it into a moral outrage that is fuel for rational oppositional action.

Transcending romantic and individualistic approaches to ecology, we may finally face the everyday questions of social and political transformation. Ecology may then begin to strive to create the political pre-conditions for establishing an ecological society. While the notion of illustrative opposition proposed in these pages offers a way to rethink such pre-conditions, it cannot replace the need to build a wider revolutionary struggle. Instead, it provides a way to broaden discussions of ecological issues to include the widest revolutionary vision possible. That vision is one of direct democracy: the passionate process through which citizens may claim the political power to create a rational, ecological, and desirable society.

An ecology of everyday life is about reaching for this desirable society, reclaiming our humanity as we reclaim our abilities to reason, discuss, and to make decisions about our own communities. It is about looking into the uncharted "wilderness" of democracy itself, that delicious, empowering, and deeply social process through which we become a truly humane expression of that nature for which we have yearned all along.

INDEX

Also published by

BLACK ROSE BOOKS

THE final volume in the Collected Works of Peter Kropotkin

EVOLUTION AND ENVIRONMENT

Peter Kropotkin, Introduction and Prefaces by George Woodcock, editor

Kropotkin the geographer had a social and political concern that transformed his interest in science into a larger ecological concern that outstripped the understanding of his contemporaries. He upheld the instinct of individuals to support one another, and acknowledged environmental influences on mutation and evolution. Whereas arguments at the time based all change on the drive for survival, Kropotkin's insight—now acknowledged by ecologists—insisted on the selective pressure of the environment and the importance of habitat. Kropotkin's vision foresaw the more interrelative and cooperative world that has become evident to us now as we approach the 21st century.

For this book, George Woodcock gathered together many not-yet-published articles written by Kropotkin during his life-long and mostly ignored scientific career. His introductions and prefaces help the reader to appreciate their revolutionary insights and put the articles in their historical context, scientifically, and politically.

George Woodcock (1912-1995)—poet, author, essayist and widely known as a literary journalist and historian—published more than 90 titles on history, biography, philosophy, poetry and literary criticism. Canada's leading 'man of letters', he had been called "a gentle anarchist in a state of grace". This final volume is dedicated to his memory.

255 pages
Paperback ISBN: 1-895431-44-1 $19.99
Hardcover ISBN: 1-895431-45-X $38.99

The *Collected Works of Peter Kropotkin* (ISSN: 1188-5807) is comprised of the following additional volumes: *Conquest of Bread; Ethics; Fields, Factories and Workshops; Fugitive Writings; Great French Revolution; In Russian and French Prisons; Memoirs of a Revolutionist; Mutual Aid; Russian Literature; Words of a Rebel;* as well as a biography, written by George Woodcock and Ivan Avakumovic, entitled *Peter Kropotkin: From Prince to Rebel.*

Black Rose Books is particularly proud to have worked with George Woodcock over the years; these eleven volumes are a testimony to his memory and to that of his great teacher and mentor, Peter Kropotkin.

BLACK ROSE BOOKS

has also published the following books of related interest

Anarchism and Ecology, *by Graham Purchase*
Beyond Boundaries, *by Barbara Noske*
Ecology of The Automobile, *by Peter Freund, George Martin*
Finding Our Way: Rethinking Eco-Feminist Politics, *by Janet Biehl*
Green Guerrillas, *by Helen Collinson, editor*
Intertwining, *by John Grande*
Murray Bookchin Reader, *by Janet Biehl and Murray Bookchin*
Nature and the Crisis of Modernity, *by Raymond Rogers*
Oceans Are Emptying, *by Raymond Rogers*
Political Ecology, *by Dimitrios Roussopoulos*
Politics of Social Ecology, *by Janet Biehl and Murray Bookchin*
Politics of Sustainable Development, *by Laurie E. Adkin*
Philosophy of Social Ecology, *by Murray Bookchin*
Race, Class, Women and the State, *by Tanya Schecter*
Rethinking Camelot, *by Noam Chomsky*
Solving History, *by Raymond Rogers*
Sustainability—The Challenge, *by L. Anders Sandberg, Sverker Sörlin, editors*
Triumph of the Market, *by Edward S. Herman*
Women Pirates, *by Ulrike Klausmann, Marion Meinzerin, Gabriel Kuhn*
Women and Religion, *by Fatmagül Berktay*

send for a free catalogue of all our titles
BLACK ROSE BOOKS
C.P. 1258, Succ. Place du Parc
Montréal, Québec
H3W 2R3 Canada

or visit our web site at: http://www.web.net/blackrosebooks

To order books in North America: (phone) 1-800-565-9523
(fax) 1-800-221-9985
In Europe: (phone) 44-0181-986-4854 (fax) 44-0181-533-5821

Printed by the workers of
MARC VEILLEUX IMPRIMEUR INC.
Boucherville, Quebec
for Black Rose Books Ltd.